SpringerBriefs in Earth System Sciences

South America and the Southern Hemisphere

Series Editors

Gerrit Lohmann
Jorge Rabassa
Justus Notholt
Lawrence A. Mysak
Vikram Unnithan

For further volumes:
http://www.springer.com/series/10032

Juan Federico Ponce · Marilén Fernández

Climatic and Environmental History of Isla de los Estados, Argentina

 Springer

Juan Federico Ponce
Centro Austral de Investigaciones
  Científicas
Ushuaia
Argentina

Marilén Fernández
Centro Austral de Investigaciones
  Científicas
Ushuaia
Argentina

ISSN 2191-589X   ISSN 2191-5903 (electronic)
ISBN 978-94-007-4362-5   ISBN 978-94-007-4363-2 (eBook)
DOI 10.1007/978-94-007-4363-2
Springer Dordrecht Heidelberg New York London

Library of Congress Control Number: 2013940304

Printed on acid-free paper

Springer is part of Springer Science+Business Media (www.springer.com)

Preface

Isla de los Estados is located in the extreme south of South America. It has been described by numerous sailors who have visited it at different times since 1616, categorizing it as one of the most inhospitable but spectacular islands in the world. It is separated by a distance of 30 km from the Mitre Peninsula by the Le Maire Strait to the southeast of the Isla Grande de Tierra del Fuego. Isla de los Estados occupies a strategically important position between the Antarctic Peninsula and the South American continent providing important paleoenvironmental information concerning the weather and impact of climate changes that have taken place since the last glaciation. Botanically, the island is situated in the extreme eastern end of the Subantarctic Evergreen Forest that constitutes the world's southernmost forest, extending down to 56°S only 800 km from the Antarctic Peninsula.

The climate changes that occurred in the far south of Patagonia and the Isla Grande de Tierra del Fuego in the recent geological past have been investigated mainly through geomorphological and paleoecological studies (Coronato et al. 2007).

The Beagle Channel (54° 53′S and between 66° 30′ and 70°W) to the south of Tierra del Fuego is an ancient tectonic valley that was completely filled with ice during the last glacial maximum (LGM, ca. 24 ka B.P.; Rabassa 2008). The ancient Beagle Glacier originated from the Darwin Mountain Range ice field, receiving tributary glaciers from the internal cirques and valleys of the mountainous branches of both margins of the range. At its maximum extent during the Last Glaciation, the ice front was located at Moat Point in the northern margin of the Beagle Channel (120 km to the west of Isla de los Estados), where many moraine arcs can be still seen. Judging by the basal ages of the peat bogs found along the Beagle Channel it may be inferred that around 14.6 ka B.P. the ice front had retreated some 100 km to the west of its point of maximum extension. The final retreat of the ice occurred at least around 10 ka B.P., when the first communities of steppe/tundra environment vegetation established themselves (Markgraf 1993a; Heusser 1989a, b).

The Quaternary deposits of the Isla de los Estados constitute an invaluable resource for the investigation of past climate events in the highest latitudes. This is particularly true in those places where the Subantarctic forest and the non-arboreal vegetation make up areas of tension that are sensitive to changing climatic parameters.

This book is the result of 10 years of scientific research made by the authors on Isla de los Estados. The research includes their doctoral thesis and many published scientific papers related to the island. Dr. J. F. Ponce arrive for first time to Isla de los Estados to carry out fieldwork activities on 2003. Dr. M. Fernández started with diatom analysis on 2006.

This book can be divided into two principal parts. The first part contains different social and natural aspects of this remote island and includes chapters on the scientific and historical background, physiography with topographical and hydrographical descriptions, climate and oceanographic circulation, vegetation and geology (including stratigraphy, structural geology, and geological history).

The second part comprises a reconstruction of the paleoenvironmental, paleoclimatic, and paleogeographic history of the island from the Last Glacial Maximum to the present, correlating with other paleoecological records from the southern part of Isla Grande de Tierra del Fuego and Patagonia. This second part also includes a geomorphological chapter with a characterization of the principal erosive glacial landforms in Isla de los Estados constructed by means of morphometric analysis, inventories, maps, paleogeographic, and glacial models, and a paleoecological chapter evaluating the palaeoenvironment and palaeoclimatic conditions that prevailed during the Late Pleistocene-Holocene times based on pollen and diatom analysis from three ^{14}C-dated peat bogs and lakes. Finally, the book concludes with a review of the island's archaeology and the relationship between the palaeoenvironmental history and human occupation of this island.

References

Coronato A, Borromei AM, Rabassa J (2007) Paleoclimas y Paleoescenarios en la Patagonia Austral y en Tierra del Fuego durante el Cuaternario. Boletín Geográfico de la Universidad Nacional del Comahue. Número especial Jornadas sobre Calentamiento Global: Neuquén, 18–28

Heusser CJ (1989a) Late quaternary vegetation and climate of Tierra del Fuego. Quat Res 31: 396–406

Heusser CJ (1989b) Polar perspective of late quaternary climates in the southern Hemisphere. Quat Res 32: 60–71

Markgraf V (1993a) Palaeoenvironments and paleoclimates in Tierra del Fuego and southernmost patagonia, South America. Palaeogeogr, Palaeoclim, Palaeoecol 102:53–68

Rabassa J (2008) Late cenozoic glaciations of patagonia and Tierra del Fuego. In: Rabassa J (ed) Late cenozoic of patagonia and Tierra del Fuego. Developments in quaternary science, vol 11. Elsevier, Amsterdam, pp 13–56

Acknowledgments

We thank Adrián Schiavini, Andrea Raya Rey, and colleagues for arranging travel and stay in Isla de los Estados. Argentine Navy and the crew of the "Sobral" ship are acknowledged for their logistical support and travel assistance to get to the island during the 2003 expedition. The authors would like to express a special thanks to the Professor Dr. Svante Bjorck (Lund University, Sweden), expedition leader of the Argentinian-Swedish Expedition to the island (2005), and also to the scientific crew members Dr. Barbara Wolfarth, Dr. Chirstian Hjort, Dr. Per Möller, Dr. Kalle Ljung, Dr. Oscar Martínez, Dr. Jorge Rabassa, Dr. Enrique Pianzola. Particularly thanks to the National Academy of Swedish Science, Lund, and Stockholm University for funding and supported the expedition. A particular thanks to Dr. J. Rabassa for all the valuable observations and suggestions made. Thank to Dr. Ana María Borromei who has made an important contribution to the knowledge of Isla de los Estados's palinology. Thanks to Dr. Nora I. Maidana, Dr. Hannelore Håkansson and Dr. Linda Ampel who has helped with the diatom analysis and taxonomy. To INGEOSUR-CONICET, Departamento de Geología, Universidad Nacional del Sur, Department of GeoBiosphere Science Center (Quaternary Sciences) (Lund University), Laboratorio de Diatomeas Continentales del Departamento de Biodiversidad y Biología Experimental de F. Cs. Exactas y Naturales (UBA) and CADIC-CONICET for having made available the required space and facilities to enable the carrying out of experimental and lab activities.

Contents

1 History, Previous Works and Physiography   1
   1.1   Geographical Location   1
   1.2   History   3
   1.3   Scientific Background   5
   1.4   Physiography   6
       1.4.1   Topographic Features   6
       1.4.2   Hydrography.   9
   References 10

2 Climate 13
   2.1   Magellanic Region 13
       2.1.1   Winds 13
       2.1.2   Temperature and Precipitation 15
   2.2   Isla de los Estados 18
       2.2.1   Temperature 18
       2.2.2   Precipitation 18
       2.2.3   Cloud 19
       2.2.4   Winds 20
       2.2.5   Electrical Storms 20
   2.3   Oceanic Circulation in the Magellanic Region 21
   References 22

3 Vegetation 25
   3.1   Introduction 25
       3.1.1   Associations of Vegetation on Isla de los Estados 26
       3.1.2   Evergreen Forest Formation. 26
       3.1.3   Scrub Formation. 27
       3.1.4   Magellanic Moorland Formation 28
       3.1.5   Meadows Formation. 30
       3.1.6   Alpine Formation 30
       3.1.7   Littoral Formation 32
       3.1.8   Maritime Tussock Formation 32

	3.1.9	Wooded Passages	33
	3.1.10	Parkland	33
	3.1.11	Rocky Promontories	34
References			34

4 Geology 35
- 4.1 Introduction 35
- 4.2 Stratigraphy 36
- 4.3 Structure 40
- 4.4 Geological History 42
- References 43

5 Glacial Geomorphology 45
- 5.1 Introduction 45
- 5.2 Beagle Channel, Isla Grande de Tierra del Fuego 47
- 5.3 Geomorphology of Isla de los Estados 47
  - 5.3.1 Periglacial Features on Isla de los Estados 48
  - 5.3.2 Glacial Morphology 50
- 5.4 Relationship Between the Structural Geology of the Island and its Principal Geomorphological Features 58
- 5.5 A Glaciation Model for Isla de los Estados 60
- References 65

6 Paleogeography 69
- 6.1 Introduction 69
- 6.2 Methodology 70
- 6.3 The Paleogeographical Model 70
- References 73

7 Palinology 75
- 7.1 Introduction 75
- 7.2 Materials and Methods 76
- 7.3 Results 81
  - 7.3.1 Modern Pollen Data 81
  - 7.3.2 Fossil Pollen Data. IDE-1 Section 82
- 7.4 Vegetation Reconstruction 83
- References 86

8 Diatom Analysis 87
- 8.1 Introduction 87
- 8.2 Field and Laboratory Methods 88
  - 8.2.1 Laguna Cascada Core (54° 45′51″S, 64° 20′20.7″W, ca. 10 m a.s.l.) 89
  - 8.2.2 Lago Galvarne Bog Core (54°44′16″S, 64° 19′37.9″W; 2 m a.s.l.) 89

	8.3	Results	89
		8.3.1	Laguna Cascada	89
		8.3.2	Lago Galvarne Bog	95
	8.4	Interpretation	98
		8.4.1	Laguna Cascada Profile	98
		8.4.2	Lago Galvarne Peat Bog: Evidence of Sea Level Increase and Paleoenvironmental Development	102
	References..	103

9 Reconstruction of Paleoenvironmental Conditions During Late Glacial and Holocene Times in Isla de los Estados and Their Correlation with the Beagle Channel and Southern Patagonia 105

9.1	Late Glacial-Holocene (18,000–11,500 cal. years B.P.)	105
9.2	Holocene (ca. 11,500 cal. years B.P.—Present)...............	109
References..	113

10 Archaeology 117

10.1	Introduction	117
10.2	Peopling of the Beagle Channel and Neighboring Islands.......	119
10.3	Archaeological Sites at Isla de los Estados	120
	10.3.1	Flinder III	121
	10.3.2	Colnett Beach.................................	121
	10.3.3	BC I (Crossley Bay)...........................	121
	10.3.4	BC II (Crossley Bay)	122
10.4	Faunal Assemblages from BC I and BC II Sites	122
10.5	Archaeology and Paleoenvironmental Scenarios in the Southern Tip of Tierra del Fuego.........................	123
References..	127

Chapter 1
History, Previous Works and Physiography

Abstract Isla de los Estados (Staaten Island) is located at the southernmost end of South America, forming part of the argentina province of Tierra del Fuego, Antarctica and South Atlantic Islands. The island, due to its geographical location, conforms a unique and sensitive area for Quaternary palaeoecological and palaeoclimatic studies giving information on the atmospheric and environmental conditions from cold-temperate high latitudes in the Southern Hemisphere. It is located at a distance of ca. 30 km southeast from Península Mitre, Isla Grande de Tierra del Fuego (Main Island of Tierra del Fuego) and has a surface area of 496 km². It is the southeastern end of the Andes range above present sea level. The topography is characteristic of terrain repeatedly glaciated during the Quaternary. At the eastern area of the island the topography is less rugged than at its central and western areas. Isla de los Estados was discovered in the year 1616. Currently, and continuously since 1977, a contingent of the Argentine Navy has been stationed in Puerto Parry. These people represent the entire population of the island together with the post on Observatorio Island (Año Nuevo Island).

Keywords Isla de los Estados • Location • History • Scientific background • Physiography • Hydrography

1.1 Geographical Location

Isla de los Estados is located in the far south of Argentina and forms part of the province of Tierra del Fuego. It is situated between the parallels of 54° 38′S and 54° 55′S and the meridians of 63° 48′W and 64° 46′W (Fig. 1.1).

It is included within the Argentine continental shelf by which it is connected to the Islas Malvinas/Falklands and the Isla Grande de Tierra del Fuego. It is separated from the latter by the Le Maire Strait, which is only 30 km in width. The island has a surface area of 520 km² (García 1986), a maximum length of 75 km from east to west, and an average width of 6 km (Niekisch and Schiavini 1998, unpublished

J. F. Ponce and M. Fernández, *Climatic and Environmental History of Isla de los Estados,* 1
Argentina, SpringerBriefs in Earth System Sciences, DOI: 10.1007/978-94-007-4363-2_1,
© The Author(s) 2014

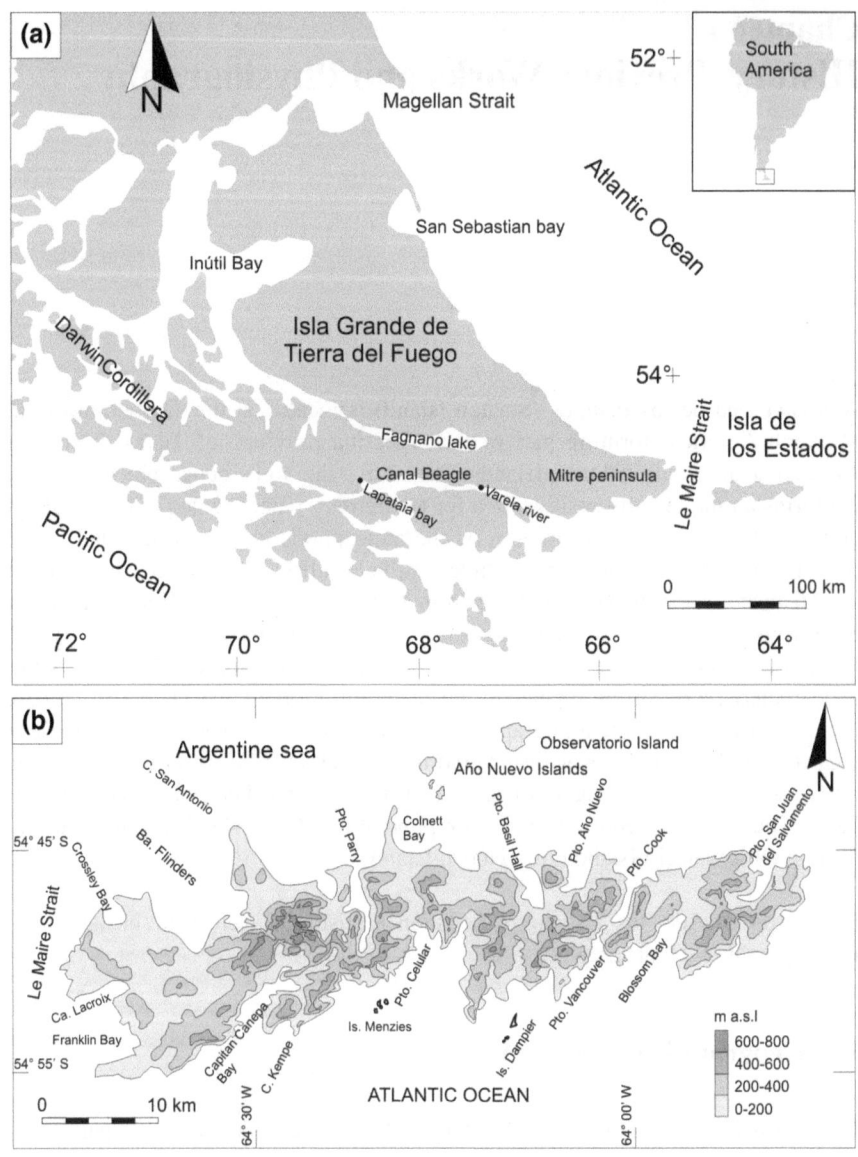

Fig. 1.1 Location of Isla de los Estados

data), with a minimum of 550 m (between Puerto Cook and Puerto Vancouver) and a maximum of 17 km (between Cabo San Antonio and Cabo Kempes) (Fig. 1.1b). It forms the extreme south eastern part of the Andes mountain range.

Isla de los Estados is surrounded by a series of small close-lying islands. To the north is the archipelago of the Año Nuevo Islands formed by a group of four islands and islets of which the largest is Observatorio Island, with an area of 400 ha (Niekisch and Schiavini 1998). Since it contains the highest point of these islands (51 m a.s.l.) a lighthouse was constructed there at the beginning of the twentieth century. A series of islets lie off the southern flank of the island the largest of which are the Menzies and Dampier islands.

1.2 History

Isla de los Estados (Staaten Island, from the Dutch *Staateneiland*) was discovered in the year 1616 by the Dutch sailors Jacques Le Maire and W. Schouten. Its name "de los Estados", refers to the Netherlands States-General, the name given in those times to the different social and political sectors of the Low Countries. The Le Maire Strait is named after the first of these explorers. Studies carried out by Chapman (1987) indicate a scarce pre Hispanic human presence in the western portion of the island.

For the following historical overview the most relevant facts have been taken from the regional history according to Goodall (1975) in Kühnemann (1976).

In 1774, Captain James Cook arrived at the Año Nuevo Islands and named the bay and the port after himself.

With the intention of establishing a sealing station Captain James Colnett disembarked at Año Nuevo Bay in the year 1787. The botanist C. Menzie, member of this expedition, collected plants in the island.

In 1828, H. Foster built an observatory in Puerto Cook and Lieutenant Kendall made a topographic map of the island. In 1833, Charles Darwin visited the island on board the ship "Beagle" during R. Fitz Roy's second South American expedition.

In the year 1884, the government of Argentina sent a "division of expedition" to the island with the task of building a lighthouse, harbour and a post for the coast guard.

Towards the end of the nineteenth century, the island housed a detachment of the coast guard (from 1884) and a maximum security prison (from 1896) in San Juan del Salvamento. In 1902 both units were moved to Puerto Cook where they remained until 1906 (Fig. 1.2). Also during this year the prison was moved to the city of Ushuaia, Tierra del Fuego.

Between the years 1892 and 1919 a meteorological observatory founded by the Argentine Navy operated on one of the Año Nuevo Islands, the "Observatorio Island" (Fig. 1.3).

In 1902 the ship "Antarctic" carrying a Swedish expedition led by Otto Nordenskjöld anchored at Año Nuevo Island on its way to Antarctica for the purpose of comparing the magnetic and meteorological instruments on board with those of the meteorological station. Underlieutenant José María Sobral, who would become the first Argentine in Antarctica and later, the first university qualified Argentine geologist, was a member of this expedition.

The Argentine government donated the island to Captain L. Piedra Buena in the year 1868 in appreciation for services rendered in the defence of the sovereignty

Fig. 1.2 Ruins of the Coast Guard station at Puerto Cook (Photo: Ponce 2005)

Fig. 1.3 Lighthouse on Observatorio Island (Photo: Ponce 2005)

over the southern territories. Piedra Buena indiscriminately exploited the natural resources of the island, operating penguin and sea-lion stations, the result of which, in addition to the actions of other hunters, was a serious blow for many of the species exploited.

Piedra Buena introduced, in 1868, a group of domesticated goats (Vinciguerra 1883 in Niekisch and Schiavini 1998) to be used for food of which there are no details regarding the original number of goats freed or the first places where this happened. In 1976 seven individuals of red deer were introduced in Crossley Bay by the National Parks Administration.

During recent decades various groups of cattle have been introduced at different times but there is no reliable information regarding the dates, numbers introduced or breeds. The brown rat was introduced at various times to Isla de los Estados in an accidental way on board the ships that visited the ports or ran aground along the coasts (Niekisch and Schiavini 1998). The Norwegian rat is found throughout the island (Niekisch and Schiavini 1998). In contrast, the black rat and European rabbit are restricted to Isla Observatorio (Massoia and Chebez 1993). Apart from beef cattle, the other species mentioned continue to spread in the island, negatively affecting the natural ecosystems.

Currently, and continuously since 1977, a contingent of the Argentine Navy has been stationed in Puerto Parry. This post is occupied by four people with rotations of 40 days. These people represent the entire population of the island together with the post on Observatorio Island (Año Nuevo Island). Isla de los Estados is without doubt one of the most isolated and uninhabited places in the Argentine territory.

1.3   Scientific Background

Since its discovery in 1616, Isla de los Estados has been visited by a great number of botanists and naturalists who collected plants and made observations on the vegetation. Moore (1983) presented a summary of the different botanic expeditions that made it to the island. Twenty-one expeditions can be counted that made collections between 1774 and 1974. The majority of these expeditions only stopped in passing and the examination they made of the flora was not intensive. Kühnemann (1976) and Dudley and Crow (1983) made interesting descriptions of the flora with a complete analysis of the different communities of vegetation present. More recently, Lavallol and Cellini (2006) produced a map of the vegetation on Isla de los Estados from the interpretation of satellite images without supporting field work.

In Isla de los Estados only three palynological studies have been completed. The oldest of these is a study done by Jhons (1981) who carried out the analysis of three peat bog sections found in Puerto Vancouver, Puerto Celular and Crossley Bay, respectively. Ponce 2009 and Ponce et al. (2011a) performed a pollen analysis of two peat bogs in Franklin Bay, the oldest of these with a basal age of 12,729 ^{14}C years BP. More recently Björck et al. (2012) presented pollen data from a peat bog located between the Lovisato and Galvarne lakes (Fig. 1.1b), with a basal age of 13,386 ^{14}C years BP.

Additionally, Unkel et al. (2008) conducted paleoclimatic studies using a geochemical analysis of a profile extracted from a peat bog located in the area around Lake Galvarne (Colnett Bay), on the northern coast of the island (Fig. 1.1b) and from specimens of lacustrine sediments taken from Lake Cascada (Fig. 1.1b). The analysis covered a time period understood to be between 16 and 10 ^{14}C ka BP.

The only study of diatoms in this island was done as part of a doctoral thesis written by Fernández (2013). In this thesis, two samples taken from two peat bogs located in the central sector of the island, Lake Galvarne peat bog, and a peat bog on the shore of Lake Cascada were analysed. Partial analyses of this diatomological study have already been published in Unkel et al. (2010) and Fernández et al. (2012).

As regards to geology, there are very few published works. The first references are of studies undertaken towards the end of the nineteenth century, such as the description made by the Italian geographer Lovisato (1983), who was part of an Italian-Argentine scientific commission that explored the island in 1881. Harrington (1943) accompanied the botanist Castellanos in 1934, making geological and pale-ontological observations. Also of note are those observations made by Harrington regarding work done by Bonarelli (1917) on the peat bogs of Tierra del Fuego, who, without visiting the island, proposed a geological sketch. Of the more recent contributions, the most interesting include those of Dalziel et al. (1974), in which important contributions are made to the stratigraphic and structural understanding of the island's geology, and Caminos and Nullo (1979), who present a complete stratigraphic, lithological, and structural description together with some geomor-phological characteristics. Recently, Ponce and Martínez (2007) revealed the dis-covery of sedimentary deposits of possibly Paleogene age on the coast of Crossley Bay (NW of Isla de los Estados). Lastly, Moller et al. (2010) made descriptions of glaciogenic sedimentary sequences in cliffs at various localities along the coasts of Colnett Bay and Crossley Bay. These authors interpreted, based on OSL and C^{14} dating, that these sequences come from the last glacial cycle.

The earliest geomorphological descriptions of Isla de los Estados are from Caminos and Nullo (1979) and García (1986). In these contributions the princi-pal geomorphological features of the island are described, dividing it into two prin-cipal regions: a central eastern mountainous region and western region of flatter morphology. Ljung and Ponce (2006) analysed periglacial features in San Juan del Salvamento in the extreme east of the island. These features were observed above the current tree line (400 m a.s.l.) at approximately 500 m a.s.l. Lastly, Ponce et al. (2010) and Ponce and Rabassa (2012) made a morphometric analysis and mapped the glacial landforms of erosive origin present on the island, generating a model of glaciation.

1.4 Physiography

1.4.1 Topographic Features

The relief of Isla de los Estados, as well as the small islands that surround it, is very irregular. The longitudinal axis of the island is marked by a mountainous line 50 km in length made up of peaks that are between 400 and 800 m a.s.l. in height. This axis gives the fallen "S" shape created by the adjustment of the orographic line to the axes of the Andean structures. The coasts are, in general, highly cut, with a large number of fjords, coves and bays. The length of the entire coastline of Isla de los Estados is in the order of 292 km (García 1986).

The terrain is very uneven, although the highest points are not very elevated (Fig. 1.4). Mount Bove is the highest point on the island at 823 m a.s.l. The hills

Fig. 1.4   View to the west at 700 m a.s.l. over San Juan del Salvamento (Photo: Ponce 2005)

Fig. 1.5   Western sector of Isla de los Estados (Photo: Ponce 2005)

are steep and pointed with sharp crests and deep valleys of glacial origin (García, 1986). Many of these valleys in their lower reaches are currently occupied by the sea thus producing the highly serrated appearance of the coastline.

In the eastern sector of Isla de los Estados the relief is less abrupt than in the centre and the west (Figs. 1.5 and 1.6). The landforms are more rounded in appearance and

Fig. 1.6 Eastern and central sector of Isla de los Estados (Photo: Ponce 2010)

Fig. 1.7 Northern coast of Isla de los Estados (Photo: Ponce 2010)

lower in height, just as they are in the Año Nuevo Islands. In the rest of the island
sharp coasts and steep hills are prevalent.

The southern coast is steeper and craggier than the northern coast (Figs. 1.7 and
1.8). The littoral composition of the southern coast includes steep-walled cliffs
next to mountains that rise to great heights only a few metres from the coast. In
this sector fjords and sheltered bays are very common in contrast to the northern
coast where, despite the presence of cliffs, these fjords and bays are rare. Beaches
with gentle slopes are very scarce and they are predominately pebbly with rounded
pebbles of greatly varying sizes. There are some rare sandy beaches that are small
in size.

Fig. 1.8 Coastal cliffs in Puerto Vancouver, southern coast of Isla de los Estados (Photo: Ponce 2005)

1.4.2 Hydrography

The mountainous line that extends in an east–west direction across Isla de los Estados defines the hydrographical watersheds, south and north, with no major differences between them (Niekisch and Schiavini 1998). The water courses are short in length, extremely fast running and with flow volumes of little significance. They are all apparently permanent owing to the abundant precipitation and the high water content of the ground, which is saturated for the majority of the year.

The streams in the central and eastern sectors are shorter in length owing to the steeper relief (Fig. 1.9). In general, they include many waterfalls and rapids near their origins. They flow to the sea over the steep walls of fjords producing elevated waterfalls and sometimes along the stepped floors of the glacial valleys.

The western sector rivers, where the relief is gentler, are greater in length, reaching in some cases 6 km long (García 1986). In this sector they flow along extensive peat bogs, low-lying and marshy areas.

None of the streams or freshwater courses has been given proper geographical names. This is because the toponomy has not been developed by inhabitants of the island, but rather by sailors (Zanola, unpublished). The ground is impermeable or highly saturated which produces a high density hydrological network. Sottini (1989) identified 80 hydrographical basins after surveying aerial photographs, but without working in the field.

Around 130 freshwater lakes have been identified. They are located within glacial valleys and cirques at different altitudes (Fig. 1.9). In some cases "rosary lakes" can be observed (series of lakes that are interconnected by small water courses) that flow out through streams of varying dimensions to the coasts of the island (Niekisch and Schiavini 1998).

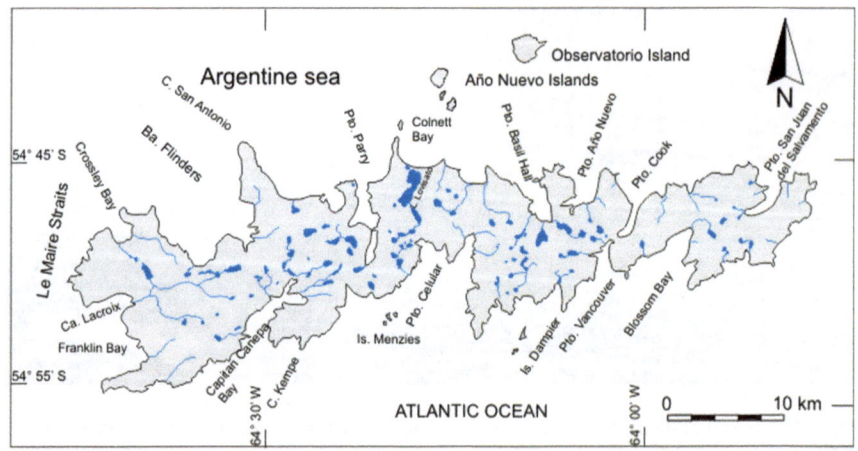

Fig. 1.9 Map of Isla de los Estados showing the distribution of the principal lakes, ponds and streams

Owing to the high humic acid content and the presence of peaty ground the surface water on the island is coloured to a reddish brown (Niekisch and Schiavini 1998).

References

Björck S, Rundgren M, Ljung K, Unkel I, Wallin A (2012) Multy-proxy analysis of a peat bog on Isla de los Estados, easternmost Tierra del Fuego: a unique record of the variable Southern Hemisphere Westerlies since the last deglaciation. Quatern Sci Rev 42:1–14

Bonarelli G (1917) Tierra del Fuego y sus turberas. Dirección Nacional de Minas y Geología XII(3):119

Caminos R, Nullo F (1979) Descripción Geológica de la Hoja 67 e, Isla de los Estados. Territorio Nacional de Tierra del Fuego, Antártida e Islas del Atlántico Sur. Servicio Geológico Nacional. Boletín 175:52

Chapman A (1987) La Isla de los Estados en la prehistoria. EUDEBA, Buenos Aires

Dalziel IWD, Caminos R, Palmer KF, Nullo FE, Casanova R (1974) South Extremity of Andes: Geology of Isla de los Estados, Argentina, Tierra del Fuego. Am Assoc Pet Geol Bull 58(12):2502–2512

Dudley TR, Crow GE (1983) A contribution to the Flora and Vegetation of Isla de los Estados (Staaten Island), Tierra del Fuego, Argentina. American Geophysical Union, Antarctic Research Series, vol 37.Washington DC, pp 1–26

Fernández M (2013) Los paleoambientes de Patagonia meridional, Tierra del Fuego e Isla de los Estados en los tiempos de las primeras ocupaciones humanas. Estudio basado en el análisis de diatomeas. Unpublished doctoral thesis, Facultad de Ciencias Naturales y Museo, Universidad Nacional de La Plata. La Plata

Fernández M, Maidana NI, Rabassa J (2012) Palaeoenvironmental conditions during the Middle Holocene at Isla de los Estados (Tierra del Fuego, 54°, Argentina) and their influence on the possibilities for human exploration. Quatern Int 256:78–87

García MC (1986–1987) Estudio de algunos rasgos geomorfológicos de la Isla de los Estados. Unpublished Graduation Thesis, Universidad Nacional del Centro de la Provincia de Buenos Aires and Centro Austral de Investigaciones Científicas, CONICET, p 53

Goodall RNP (1975) El primer blanco fueguino. In: Juan E, Belza, En la Isla del Fuego, 2ª Colonización. Instituto de Investigaciones Históricas de Tierra del Fuego, Buenos Aires, p 366

Harrington HJ (1943) Observaciones geológicas en la Isla de los Estados. Anales Museo Argentino de Ciencias Naturales. Geología 29:29–52

Jhons WH (1981) The vegetation history and paleoclimatology for the Late Quaternary of Isla de los Estados, Argentina. Unpublished Ph.D. Dissertation, Michigan State University, USA

Kühnemann O (1976) Observaciones ecológicas sobre la vegetación marina y terrestre de la Isla de los Estados (Tierra del Fuego, Argentina). Ecosur, Argentina 3(6):121–248

Lavallol CJ, Cellini JM (2006) Informe técnico: "Mapa de vegetación de la Isla de los Estados". Programa de investigación Geográfico Político Patagónico, Instituto de Ciencias Políticas y Relaciones Internacionales. Grupo Ambiental Patagónico. Facultad de Ciencias Fisicomatemáticas e Ingeniería, Universidad Católica Argentina, 2 mapas + informe, p 17

Ljung K., Ponce, J. F. 2006. Periglacial features on Isla de los Estados, Tierra del Fuego, Argentina. III Congreso Argentino de Cuaternario y Geomorfología. Actas, vol 1. Córdoba, pp 85–90

Lovisato D (1983) Una escurzione geologica nella Patagonia Terra del Fuoco. Bul Societa Geografica Italiana II(8):(5 and 6)

Massoia E, Chebez JG (1993) Mamíferos silvestres del Archipiélago Fueguino. Literatura of Latin America, Buenos Aires, p 261

Moller P, Hjort C, Björck S, Rabassa J, Ponce JF (2010) Glaciation history of Isla de los Estados, southeasternmost south America. Quatern Res 73(3):521–534

Moore MD (1983) Flora of Tierra del Fuego. Antony Nelson England, Missouri Botanical Garden, St.Louis, p 369

Niekisch M, Schiavini A (1998) Desarrollo y conservación de la Isla de los Estados. Unpublished technical report, CADIC, Ushuaia, p 70

Ponce JF (2009) Palinología y geomorfología del Cenozoico tardío de la Isla de los Estados. Unpublished Doctoral Thesis, Universidad Nacional del Sur, Bahía Blanca, Argentina, p 191

Ponce JF, Martínez O (2007) Hallazgo de depósitos sedimentarios postcretácicos en Bahía Crossley, Isla de los Estados, Tierra del Fuego. Revista de la Asociación Geológica Argentina 62(3):467–470

Ponce JF, Rabassa JO, Martínez O (2010) Fiordos en Isla de los Estados: Descripción morfométrica y génesis de los únicos fiordos en la Patagonia Argentina. Revista de la Asociación Geológica Argentina 65(4):638–647

Ponce JF, Borromei AM, Rabassa JO, Martínez O (2011a) Late Quaternary palaeoenvironmental change in western Staaten Island (54.5°S, 64°W), Fuegian Archipelago. Quater Int 233:89–100

Ponce JF, Rabassa J (2012) Geomorfología glacial de la Isla de los Estados, Tierra del Fuego, Argentina. Revista de la Sociedad Geológica de España 25(1–2):67–84

Sottini R (1989) Cuencas hídricas de la Isla de los Estados. In: Iturraspe R, Sottini R, Schroeder C, Escobar J (eds) Grupo de Hidrología: Hidrología y variables climáticas del Territorio de Tierra del Fuego. Información Básica. Contribución científica. CADIC 7. Ushuaia, p 84

Unkel I, Björck S, Wohlfarth B (2008) Deglacial environmental changes on Isla de los Estados (54.4°S), southeastern Tierra del Fuego. Quatern Sci Rev 27:1541–1554

Unkel I, Fernández M, Björck S, Kjung K, Wolfarth B (2010) Records of environmental changes during the Holocene from Isla de los Estados (54.4° S), southeastern Tierra del Fuego. Global Planet Change 74:99–113

Chapter 2
Climate

Abstract The climate of Isla de los Estados is strongly influence by the centre of the sub-polar low pressure which develops around the Antarctic Circle. The current climate of Isla de los Estados is cold and humid and corresponds to general classification of Oceanic Insular Cold Climate. Summer has a mean temperature of 8.3 °C, with mean daily extremes of 16.2 and 3 °C. Winter mean temperature is 3.3 °C, with mean daily extremes of 7.4 and −4 °C. Though no reliable records are yet available, rainfall is estimated to be in the range of 2,000 mm/year, but actual precipitation may be highly variable across the island, particularly in altitude. Prevailing winds are from the southwest and the northwest and they are active throughout the year. Isla de los Estados is washed by the western branch of the Malvinas/Falkland current, an arm of the Antarctic Circumpolar Current that brings cold Sub-Antarctic waters.

Keywords Isla de los estados • Magellanice region • climate • Sub-polar low pressure • Antarctic circumpolar current

2.1 Magellanic Region

2.1.1 Winds

The Magellanic Region (Tierra del Fuego and adjacent regions of Patagonia south of 51°S) is situated between the southern edge of the semi-permanent subtropical high pressure cell, the direct influence of which extends up to around 40°S throughout the year, and the centre of the sub-polar low pressure which develops around the Antarctic Circle (Fig. 2.1). These pressure systems have only small seasonal variations and change little in intensity, and the Westerlies prevail in this region all year round (Prohaska 1976; Burgos 1985; Endlicher and Santana 1988; in Tuhkanen 1992).

J. F. Ponce and M. Fernández, *Climatic and Environmental History of Isla de los Estados, Argentina*, SpringerBriefs in Earth System Sciences, DOI: 10.1007/978-94-007-4363-2_2, © The Author(s) 2014

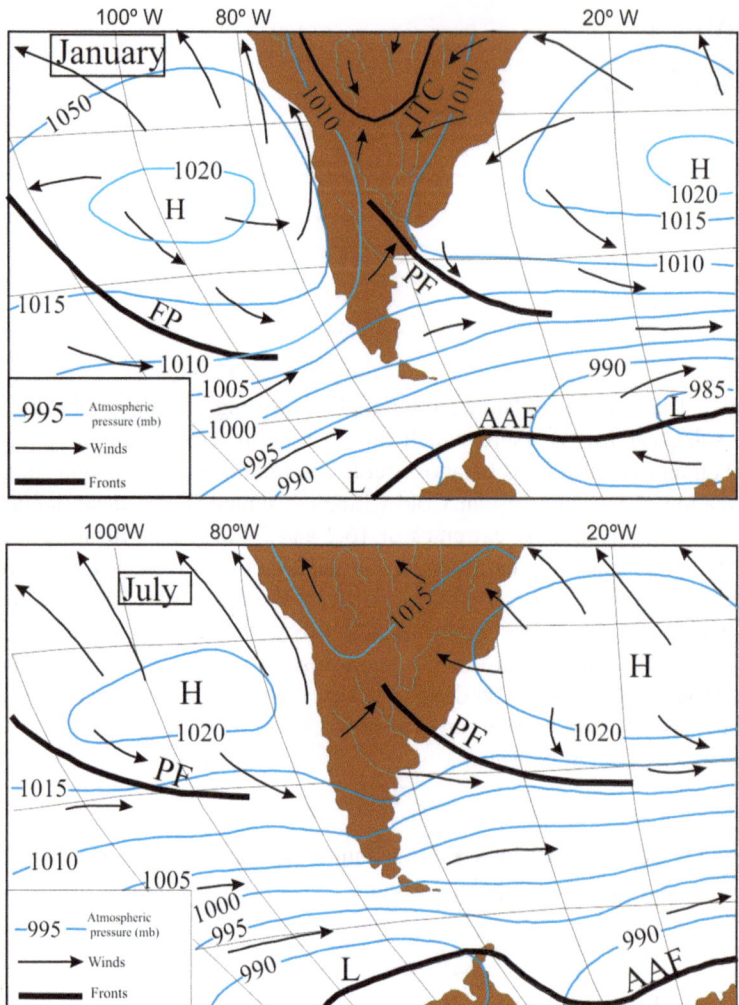

Fig. 2.1 The Magellanic Region is located between the southern edge of the semi-permanent subtropical high pressure cell (H), and the centre of sub-polar low pressure that occurs approximately on the Antarctic *Circle* (L). *ITC* inter-tropical convergence, *PF* polar front, *AF* Antarctic Front. Modified from Tuhkanen (1992)

In the western sector of the Drake Passage north-easterly components slightly dominate. Towards the east of the Passage the westerly and south-westerly components increase such that to the southeast and east of Tierra del Fuego the most frequent wind direction is easterly, with north-easterly and south-easterly being almost equally frequent. The constant south-easterly wind direction in the region of Ushuaia is attributed to diversions caused by relief. The orography also contributes to the existence of a regime of calms significantly higher than that in the steppe, something that is seen most especially during winter. The winds are in general stronger and more persistent during spring and summer (Tuhkanen 1992).

The average annual wind velocity in the west and southwest of Tierra del Fuego is 12 ms^{-1}, with a maximum that exceeds 30 ms^{-1} for the whole month. Leeward of the mountain range, the obstacle effect reduces the mean annual wind velocity to around 4–6 ms^{-1} (e.g., the city of Punta Arenas, Chile). The effect decreases with increasing distance from the mountains. On the Patagonian coast, the wind velocity reaches an average of 8 ms^{-1} (e.g., Río Grande 8.3 ms^{-1}) (Tuhkanen 1992).

The temporary variations in the directions and intensity of the winds are mainly associated with the passage of cyclonic depressions towards the east and southeast following the line of the continent (Arnett 1958).

The cyclones are formed in the polar front of the South Pacific which occurs between 110° and 120° West, between the two cells of the high pressure belts of the South Pacific. (Taljaard 1972).

The anticyclones manage to cross the continent without major difficulty. The cyclonic depressions, by contrast, disintegrate as they approach the continent and are regenerated to its west. A high percentage of the centres of low pressure cross the South American sector in latitudes south of Tierra del Fuego generating the famous storms of the Drake Passage (Servicio Meteorológico Nacional 1994).

Tierra del Fuego and the south of Patagonia are occasionally, and in particular during the winter, under the influence of stable cold Antarctic air when a line of high pressure develops behind a series of cyclones. This results in a current originating in the Antarctic causing brief sunny periods that are dry, cold and usually windy. Often this Antarctic air is warmed on its journey north and humidified by the surface of the sea. In this way it reaches the region as a damp and unstable current that afterwards joins the western circulation (Zamora and Santana 1979; Burgos 1985; Endlicher and Santana 1988).

The winds coming from the west or "West Wind Drift" that come into the region from the west-southwest exercise a pronounced oceanic trend in winter all along the Pacific Ocean coast. No continuous ice-sheets are seen in the Beagle Channel. The direct oceanic influence penetrates from the east through the Magellan Strait, separating two "cold nuclei" in the south of Patagonia and in the interior of Tierra del Fuego (Tuhkanen 1992).

2.1.2 Temperature and Precipitation

The influence of the Pacific Ocean is more evident in the temperatures during winter. A marked difference is seen between the temperatures on the east and west coasts of Tierra del Fuego. The average temperature of the Pacific coast reflects the temperature of the ocean, which is 4 °C in winter and spring. In particular, the entire coastal area shows average temperatures above freezing point in the coldest months whereas the areas with high mountains are exposed to freezing conditions. In the more inland areas of the island, probably not too far from the coast, the

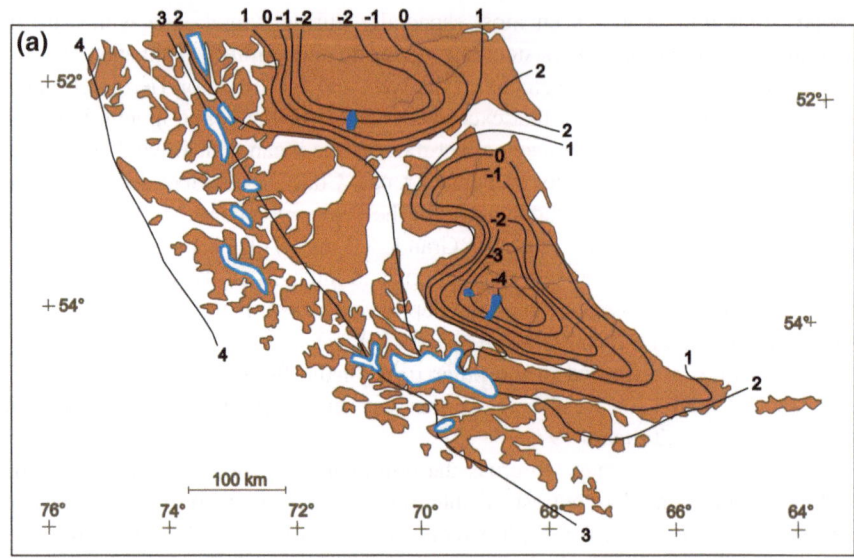

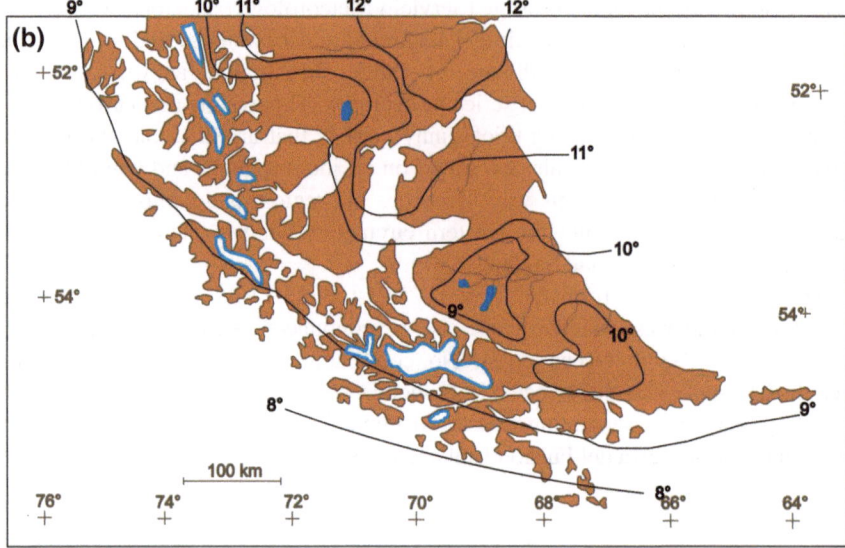

Fig. 2.2 a Winter isotherms for the Magellanic Region. Modified from Tuhkanen (1992).
b Summer isotherms for the Magellanic Region. Modified from Tuhkanen (1992)

average temperature during winter is below freezing in the lowlands (Fig. 2.2a).
The temperature gradients around the "cold centres" are relatively steep, but the
north–south direction gradients are not evident.

The temperature conditions in summer show a slight continental influence
from the Pampas in the north (Fig. 2.2b). The southeast margin of the Fuegian

archipelago has the coldest summers (8–9 °C in the hottest months), reflecting the oceanic temperature, which is 7 °C in summer. The average temperatures in the warmest months of the year show a clear gradient from the west to east following the oceanic gradient and also from north to south, which partially coincides with the oceanic gradient (Tuhkanen 1992).

As regards the vertical temperature gradient, studies carried out by Iturraspe et al. (1989) indicate an average value of 0.55 °C/100 m for the region (0.7 °C in December and 0.4 °C in July).

Due to the forced orographic ascent of the maritime air masses on the western slopes of the mountain range, very high precipitation is produced in the islands at the foot of mountains in the order of 4,000 mm per year, up to approximately 53°S (Fig. 2.3) (Servicio Meteorológico Nacional 1994). Beyond this latitude the mountain range definitively changes its N–S direction to a more W–E one, so reducing its influence on the general flow. Consequently, the amount of annual precipitation falls dramatically to 2,000 mm, around Santa Inés Island at 54°S and 1,000 mm annually on Hoste Island (Servicio Meteorológico Nacional 1994).

Between 69° and 68° 34′W an appreciable change is seen in the height of the mountains generating a regional leeward effect which manifests itself in the reduction of precipitation to 500 mm in the Beagle Channel and Navarino Island. The said effect loses intensity in the far west of Tierra del Fuego where the precipitation increases to 1,000 mm per year and it is non-existent on Isla de los Estados in whose eastern sector 1,400 mm of precipitation are registered annually (Servicio Meteorológico Nacional, unpublished, period 1982–1986).

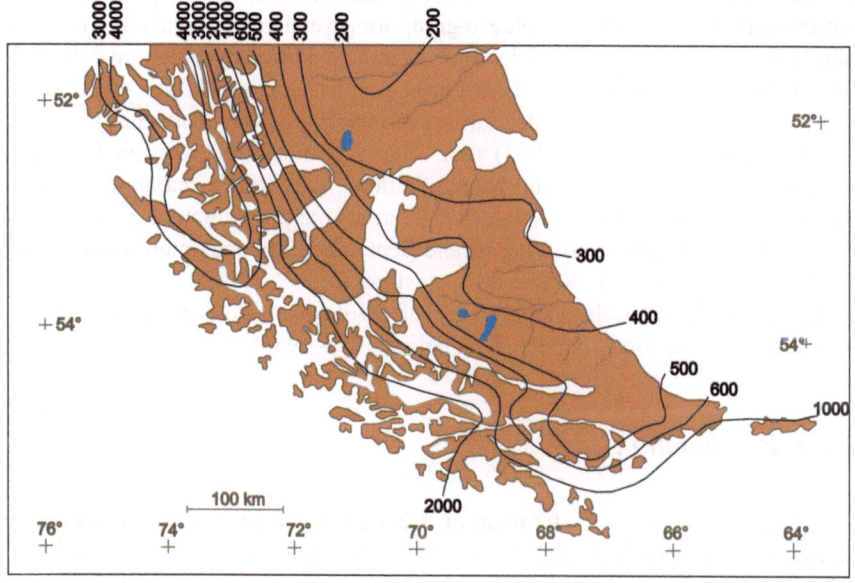

Fig. 2.3 Annual precipitation (mm) for the Magellanic Region. Modified from Tuhkanen (1992)

2.2 Isla de los Estados

The climate of Isla de los Estados is cold and wet. The numerous and rapid daily variations in the weather conditions create a constant situation of instability just as in the rest of Tierra del Fuego. Following the climatic classification of Köppen (1936) it may be classified as a Ef climate, that is to say a cold climate in which the mean temperature of the warmest month is below 10 °C and wet during the whole year. According to García (1986), this climate corresponds to the general classification of Cold Insular Oceanic. When considering the climatic scale of Knoche and Borzacov (1947), two types of climate can be deduced: "Moderate Cold" if you consider the average annual temperature, but taking into account the absolute maxima and minima, classifies it as "Temperate" and "Cold" respectively.

2.2.1 Temperature

In Isla de los Estados the air temperatures are low, but without extreme minima. In contrast to the west of the archipelago of Tierra del Fuego, the island has greater seasonal differentiation. In summer, the mean temperature is 8.3 °C, with mean extremes of 16.2 and 3.0 °C. The mean temperature in winter is 3.3 °C with mean extremes of 7.4 and −4 °C. This average is less than in Tierra del Fuego although the absolute minima are more moderate due to the influence of the ocean. The seawater has a mean temperature that varies between 5 and 7 °C (Dudley and Crow 1983).

The Argentine Meteorological Service maintained a meteorological station on Observatorio Island in the Año Nuevo group for several years and another in Puerto San Juan del Salvamento (Fig. 1.1). Their average temperature values are 6–8 °C for both meteorological stations for the warmest month (January) and 0 to −2 °C in Año Nuevo and 2–4 °C in San Juan del Salvamento for the coldest month (July).

El Derrotero Argentino (1962) (i.e., The Argentine Log) registered different temperature values for the "Año Nuevo Station", indicating a maximum mean of 11 °C for summer (December, January, February) and a minimum mean of 1 °C for the months of June, July, August and September. The maximum extremes are 16–18 °C in summer and −6 to −8 °C in winter.

Isla de los Estados has an average of 30 days of frost per year (Servicio Meteorológico Nacional, unpublished, period 1982–1986).

2.2.2 Precipitation

The precipitations are very frequent on Isla de los Estados but the absolute total values are not very high. De Fina (1972) indicates values of precipitation of 200–350 mm and 100–200 mm for the meteorological stations of San Juan de Salvamento and Año Nuevo Islands respectively during the summer months. In

Fig. 2.4 Snow in Colnett Bay (Photo: J. F. Ponce 2005)

the winter season the values are 350–500 mm for San Juan del Salvamento and 100–200 mm on the Año Nuevo Islands.

A map made by L.J. Scabella included in De Fina (1972) showed the isohyets, expressed in millimetres per year, of the mean annual precipitation. On this map, the 500 mm isohyet corresponds with the western extreme of Tierra del Fuego, whereas Isla de los Estados is included in the 1000 m isohyet.

Dudley and Crow (1983) showed that the annual mean of numbers of days with precipitation was 251.5 days with 1,447 mm, those of greater importance occurring in the winter months. June was the wettest month and October the driest.

The greatest frequency of snow in the region is claimed by Isla de los Estados with 60 days with snow per year (Fig. 2.4). The greatest number of days with snowfall frequently occur during the winter months, with a mean of 33 days (Dudley and Crow 1983). Autumn and spring also have precipitation in the form of snow but to a lesser degree. Apart from the months of May and June, in the lower lying areas the snow probably does not persist on the ground for long periods.

2.2.3 Cloud

The amount of cloud is high in this zone, occasionally a covering of cloud virtually sits on Isla de los Estados, obscuring the hilltops (Fig. 2.5). The estimation of annual cloudy days is 74 %. Although these conditions are quite well distributed throughout the year, they reach a maximum of 80 % in June and a minimum of 68 % in October (Dudley and Crow 1983). The visibility in general terms is good and there is usually little fog. An average of 16 days with fog are registered annually.

Fig. 2.5 Low cloud on the southern edge of Lake Lovisato, central area of Isla de los Estados
(Photo: J. F. Ponce 2005)

The relative humidity is very high and remains fairly uniform throughout the
year, with a maximum of 87 % in June and a minimum of 76 % in December
and January. The weather patterns change quickly and unpredictably in this region
(Dudley and Crow 1983).

2.2.4 Winds

The winds are constant and strong, predominantly from the northwest and southwest (Kühnemann 1976). They are heavily laden with humidity. The storms are
strong (strength greater than 8 for 73 days a year) and frequent (Fig. 2.6). The
greatest intensity of wind is achieved in winter with an average of 37 km/h for the
month of August and 24 km/h in December (Dudley and Crow 1983).

2.2.5 Electrical Storms

This type of phenomenon is not very common in the Tierra del Fuego and
South Atlantic Islands zone. In the majority of the stations that the National
Meteorological Service possesses, less than 10 days with electrical storms
were registered in 10 years, with the exception of Isla de los Estados and the
Evangelistas islet that registered 15–20 days each 10 years respectively (Servicio
Meteorológico Nacional, unpublished, period 1982–1986).

Fig. 2.6   Storm in Colnett Bay, December 2005 (Photo: J. F. Ponce 2005)

2.3   Oceanic Circulation in the Magellanic Region

South America is situated to the north of the Southern Ocean, a circumpolar body of open water from 20° to 30° of latitude in width, occupying the space between the Antarctic continent and the Atlantic, Pacific and Indian Oceans (Hamon and Godfrey 1978). Interrupted only by small islands beyond 50°S, the Southern Ocean is one of the largest and most remote regions of the planet. Its waters, located between 40° and 60° of latitude South, travel predominantly eastwards in the "West Wind Drift", or Antarctic Circumpolar Current, controlled by the strong southern Westerlies. Along the coast of Antarctica, where the Polar drift of the Easterlies ("East Wind Drift") is found, the Antarctic Circumpolar Current moves from east to west. The Humboldt current separates from the "West Wind Drift" and travels towards the Equator along the coast of Chile, producing Sub-Antarctic conditions around 48°S across the Pacific and a cold maritime climate in the low subtropical latitudes (Heusser 2003). A second arm of the Antarctic Circumpolar Current continues towards the west where it is known as the Cape Horn Current (Fig. 2.7) travelling across the south of the Isla Grande de Tierra del Fuego. This current turns towards the north, stretched by the point of South America and the islands of the Scotia Arc and divides into two arms, one passes either side of the Malvinas/Falkland Islands becoming the Malvinas/Falkland Current and the other goes around the South Georgia Islands. Accordingly, Isla de los Estados is washed by the western branch of the Falkland current that brings cold Sub-Antarctic waters. It is a fairly strong current having a mean velocity of 15−20 knots (Kühnemann 1976).

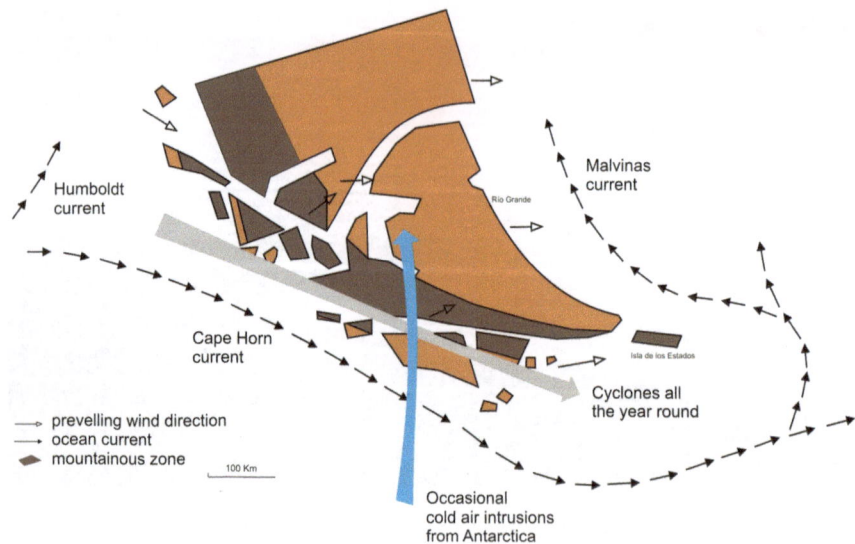

Fig. 2.7 Principal factors that affect the climate in Tierra del Fuego. Modified from Tuhkanen (1992)

According to Deacon (1937, 1960, 1963), the Antarctic Convergence or Polar Oceanic Front is located at around 50°S on average, where the temperature of the surface water falls rapidly towards the pole approximately 2 °C. The Subtropical Convergence is located at 40°S, and here the water temperature falls 4 °C. This front is considered to represent the northern limit of the Southern Ocean. The Antarctic Convergence (Gordon 1967; Gordon and Goldberg 1970) provides a reference to outline the geographical field of the Antarctic and Sub-Antarctic.

References

Arnett JS (1958) Principal tracks of southern hemisphere extratropical cyclones. Monthly Weather Rev 86:41–44

Burgos JJ (1985) Clima del extremo sur de Sudamérica. In: Boelcke O, Moore DM, Roig FA (eds) Transecta botánica de la patagonia austral. CONICET, Instituto de la Patagonia and Royal Society, pp 10–40

Bagnati RH Derrotero AP III (1962) Archipiélago Fueguino Islas Malvinas 3° (ed) I. Servicio de Hidrografía Naval p 324

Deacon GER (1937) The hydrology of the Southern Ocean. Discov Rep 15:1–24

Deacon GER (1960) The southern cold temperate zone. In: Proceedings of the Royal Society. Series B 152:441–447

Deacon GER (1963) The Southern Ocean. In: Hill MN (ed) The Sea, Interscience Publisher, London, 2:281–296

De Fina AL (1972) El clima de la región de los bosques Andino-patagónicos argentinos. In: Dimitri M (ed) La región de los bosques andino-patagónicos. Sinopsis general. Colección científica del INTA (Instituto Nacional de Tecnología Agropecuaria) Buenos Aires 10:35–58

Dudley TR, Crow GE (1983) A contribution to the Flora and Vegetation of Isla de los Estados (Staaten Island), Tierra del Fuego, American Geophysical Union, Antarctic Research Series, Argentina, vol 37, Washington DC, pp 1–26

Endlicher W, SantanaA (1988) El clima del sur de la Patagonia y sus aspectos ecológicos. Un siglo de mediciones climatológicas en Punta arenas. Anales del Instituto de la Patagonia, Serie ciencias naturales Punta Arenas, Chile 18:57 86

García MC (1986–87), Estudio de algunos rasgos geomorfológicos de la Isla de los Estados. Unpublished graduation thesis, Universidad Nacional del Centro de la Provincia de Buenos Aires and Centro Austral de Investigaciones Científicas, CONICET, Ushuaia, p 55

Gordon AL (1967) Structure of Antarctic waters between 20° W and 170° W. Antarctic Map Folio Series, USA

Gordon AL, Goldberg RD (1970) Circumpolar characteristic of Antarctic waters Antarctic Map Folio Series. 18, American Geographical Society of New York, New York, 13:1–19

Hamon BV, Godfrey JS (1978) Role of the oceans. In: Pittock AB, Frakes LA, Jenssen D, Peterson JA, Zillman JW (eds) Change and variability: a southern perspective. Cambridge University Press, UK, pp 31–52

Heusser CJ (2003) Ice age southern andes. A chronicle of paleoecological events. Developments in quaternary science 3. series editor: Jim rose. Elsevier, 5–10

Iturraspe R, Sottini R, Shroder C, Escobar J (1989) Generación de información hidro-climática en tierra del fuego. In: Hidrología y variables climáticas en Tierra del Fuego. Inf Básica Contr Cient 7. CADIC, Ushuaia, pp 4–170

Knoche W, Borzacov J (1947) Provincias Climáticas de la Argentina. In: Geografía de la República Argentina (GAEA), Buenos Aires 6:140–174

Köppen W (1936) Das geographische Sistem der Klimate. In: Köppen W, Geiger R (eds) Hadbuch der klimatologie 1C verlag von gebruder borntraeger, Berlin, pp 1–44

Kühnemann O (1976) Observaciones ecológicas sobre la vegetación marina y terrestre de la Isla de los estados (Tierra del Fuego, Argentina). Ecosur, Argentina 3(6):121–248

Prohaska F (1976) The climate of argentina paraguay and uruguay. In: Schwerdtfeger W (ed) Climate of central and south america. World survey of climatology. Elsevier, Amsterdam, 12:13–112

Nacional SM (1982–1986) Unpublished reports, CADIC, Ushuaia, p 87

Nacional SM (1994) El archipiélago de tierra del fuego. Unpublished report, Ushuaia, p 21

Taljaard JJ (1972) Synoptic meteorology of the southern hemisphere. Meteorol Monogr 13:139–213

Tuhkanen S (1992) The climate of Tierra del Fuego from a vegetation geographical point of view and its ecoclimatic counterparts elsewhere. Acta Botanica Fennica, 125:4–17

Zamora E, Santana A (1979) Oscilaciones y tendencias térmicas en Punta Arenas entre 1888 Y 1979. Anales del Instituto de la Patagonia, Punta Arenas, Chile 10:145–154

Chapter 3
Vegetation

Abstract In the island, seven vegetation types related to altitude and terrain forms are recognized. The more protected and lower mountain slopes and valleys show the development of *Nothofagus betuloides* and *Drimys winteri* forest characteristics of the true Subantarctic Evergreen Forest. In those sites constantly exposed to continuously strong wind conditions, the main vegetation is the Magellanic Moorland Formation, a mosaic of interfingered and superimposed subunits that may cover a rather small area forming blanket-like patches. The Scrub Formation occurs on mountain slopes exposed to the prevailing southwesterly and westerly winds where trees of *Nothofagus antarctica* and shrubs grow low and tortuous. Above approximately 450 m the so-called Alpine Formation occurs with sparse vegetation cover, and often includes dwarfed forms of *Nothofagus antarctica* and *Empetrum rubrum*. Soil conditions also influence the vegetation that characterizes the island. For example, the cold and damp climate favors peat development mainly at low and intermediate elevations. However, in spite of the constant soil humidity, the topography influences drainage patterns and this is reflected by the composition of the vegetation. Littoral and Maritime Tussock Formations develop along the stony and rocky coastal areas, small peats and soils above the high tide line and in rock crevices at the intertidal zone.

Keywords Isla de los Estados • Vegetation • Subantarctic evergreen forest • Magellanic Moorland • Peat bogs

3.1 Introduction

Several botanists, both specialists and collectors, have visited Isla de los Estados, but practically none has performed a thorough study of the vegetation. Nevertheless, some have included vegetation in their phytogeographical maps.

According to Cabrera (1976), Isla de los Estados corresponds to the Sub-Antarctic dominion, characterized by the presence of exuberant vegetation. Niekisch and Schiavini (1998 unpublished) considered that the forests correspond to the

J. F. Ponce and M. Fernández, *Climatic and Environmental History of Isla de los Estados,*   25
Argentina, SpringerBriefs in Earth System Sciences, DOI: 10.1007/978-94-007-4363-2_3,
© The Author(s) 2014

Magellanic district of the Patagonian Phytogeographic Province. This region constitutes, along with the southern coasts of Isla Grande de Tierra del Fuego, the only
location in Argentina where forests develop right from sea-level.

The majority of the slopes that come down from the mountains and valleys are
predominantly colonized by trees of the dominant genus *Nothofagus* and the less
abundant *Drimys* that are characteristic of the true Magellanic Evergreen Forest
(Young 1973). The places most exposed to the strong and constant winds on the
island are similar in terms of flora to the Magellanic moorland of the Fuegian
Archipelago, which is very characteristic of the Sub-Antarctic region (Dudley and
Crow 1983). In the places most exposed to the strong south-easterly and easterly
winds it is possible to observe the influence of the winds on some trees that have
bent and knotty trunks that grow in the direction of the dominant winds. The floor
of the forest is dominated by diverse species of lichens that form a very spongy
covering with high water retention.

3.1.1 Associations of Vegetation on Isla de los Estados

Seven formations or associations of vegetation have been recognized on Isla de los
Estados (Dudley and Crow 1983), which are listed as follows:

3.1.2 Evergreen Forest Formation

3.1.2.1 *Nothofagus betuloides-Drimys winteri* Association

The evergreen forest is dominated by the tree species *Nothofagus betuloides* and
Drimys winteri. The trees of *N.betuloides* stand between 8 and 15 m in height, producing a branching foliage that forms a flattened top that can reach 6 m in width
and so restrict the penetration of sunlight (Fig. 3.1). *Drimys winteri* is of a smaller
size (between 6 and 8 m tall) and has a conical or cylindrical shape starting from 1.5
to 2.5 m from the ground surface. The base stratum includes bushy and herbaceous
species such as *Gunnera magellanica, Luzuriaga marginata, Berberis ilicifolia* and
Senecio acanthifolius (especially in low-lying and damp areas), accompanied by
bryophytes and ferns such as *Hymenophyllum tortuosum* and *H. dentatum*.

In the northeastern part of the island, particularly to the west of Mount Spegazzini
the Evergreen Forest is characteristic of the Transitional Mesic Evergreen Forest as
described by Young (1973). This forest of the northeast is atypical of the Magellanic
Evergreen Forest and has an impoverished basal stratum due to the effects of dry
winds and the damage caused by goats introduced to the island. The ferns and bryophytes are scarce. Similarly, the angiosperms include impoverished elements of
Lebetanthus myrsinites and *Berberis ilicifolia*, with *Senecio smithii, S acanthifolius*
and *Gunnera magellanica* appearing sporadically in flat, muddy areas at the base of
tree-covered slopes.

Fig. 3.1 Forest of *Nothofagus betuloides* in the area near Puerto San Juan del Salvamento (Photo: Ponce 2005)

In contrast, the Evergreen Forest in the centre, east and southeast of the island is much wetter and has a forest floor that supports and exuberant development of bryophytes and terrestrial and epiphytic ferns, particularly *Hymenophyllum tortuosum*.

Lebetanthus myrsinites is a creeping bush that thrives in this area densely growing on the forest floor and often climbing around the bases of the trees. The epiphytic ferns, such as *Grammitis magellanica* and *Serpyllopsis caespitose*, and bryophytes are also abundant, as are pteridophytes that live on the ground (*Blechnum magellanicum*). The forest in this part of the island is more characteristic of the genuine Magellanic Evergreen Rain Forest that dominates the southwest coast of Chile (Young 1973).

3.1.3 Scrub Formation

Two well defined associations of vegetation are recognized within this formation.

3.1.3.1 *Nothofagus Antarctica* Association

The Deciduous Forest of the Magellanic Region is represented by *Nothofagus antarctica* from the more elevated zone. This zone is restricted to a narrow belt of vegetation in small thickets, twisted and contorted on the highest parts of the mountains

Fig. 3.2 Stunted forest of *Nothofagus antarctica* at 400 m a.s..l. in San Juan del Salvamento (Photo: Ljung 2005)

above the Evergreen Forest and just below the alpine zone (Fig. 3.2). This stunted, dense and bent vegetation of *Nothofagus Antarctica*, which is flattened towards the top is called "krummholz". It reaches a height of 1.0–1.5 m and is almost impenetrable. Occasionally impoverished specimens of *Chiliotrichum diffusum*, *Pernettya mucronata* and *Berberis ilicifolia* may be found at the base of the tangled stems of *Nothofagus* sp., which sometimes can be covered with *Hymenophyllum tortuosum* and *Sphagnum* sp. Very occasionally, small groups of *Marisppospermum grandiflorum* can be present, particularly in damp peaty areas. In the lower elevations these "krummholz" can be interfingered with vegetation associations of the Magellanic moorland and the meadows of *Marisppospermum* sp.

3.1.4 Magellanic Moorland Formation

The vegetal communities of this formation are more difficult to analyse and define since they consist of a mosaic of interwoven and overlapping subunits which can cover a fairly small or fairly extensive area forming coverings, as in the case of *Astelia pumila*. Three subunits or associations of Magellanic Moorland are recognized.

3.1.4.1 Empetrum rubrum Association

This association is the predominant subunit of the Magellanic Moorland and is principally dominated by *Empetrum rubrum*, *Pernettya mucronata* and *Marsippospermum grandiflorum.* Nothofagus betuloides and Drimys winteri appear as accompanying species. These are dispersed and have bushy, scrub-like stunted appearance. *Chiliotrichum diffusum* and *Berberis ilicifolia* are generally present with a stunted aspect. *Sphagnum*, other bryophytes and some ferns appear forming dense mounds around the lowest branches of the bushy plants, especially *Empetrum*. Although *Sphagnum* is quite common throughout this formation, genuine peat bogs of *Sphagnum* were not found anywhere on the island. Within the moorland, *Gunnera magellanica, Blechnum penna-marina* and *Luzuriaga marginata* are found associated with *Empetrum*. In the lower elevations, *Blechnum magellanicum* is present on the drier slopes of the moorland. Low and flat patches of *Rubus geoides* are often seen in the driest areas, particularly on the high, exposed promontories.

3.1.4.2 *Caltha* Association

In these places within the moorland where *Empetrum* and *Marsippospermum* are less dense, a subunit or association develops in which *Caltha dionaeifolia* and *C. appendiculata* predominate. *Myrteola nummularia, Astelia pumila, Gaultheria antarctica, Pernettya pumila, Gunnera lobata, Gaimardia australis, Tribeles australis, Perezia magellanica* and *Drapetes muscosoides* also accompany this association as secondary elements. Both species of *Caltha* mix with *Astelia pumila* often forming large patches that are uniform, dense, low-lying and flat. *Nanodea muscosa* is usually abundant in these coverings of *Caltha*.

3.1.4.3 *Astelia pumila* Association

This association develops on the higher and steeper slopes of the elevated zones where extensive coverings of low-lying, flat and firm *Astelia pumila* can be seen (Fig. 3.3). They can consist almost exclusively of *Astelia*, but other species such as *Gaimardia australis, Gunnera lobata, Pernettya pumila, Drosera uniflora, Oreobolus obtusangulus, Caltha dionaeifolia, C. appendiculata, Abrotanella emarginata, Azorella lycopodioides* and *Bolax gummifera* can also appear. *Nanodea muscosa* occurs abundantly accompanying these secondary species, but does not do so in the coverings of pure *Astelia pumila*. The rare fern *Gleichenia cryptocarpa* is found almost exclusively in these coverings. In very wet areas, particularly in zones of melting snow *Drosera uniflora* forms in patches of dense plants that give the slopes a purple colour.

Locally gradation is observed between the associations of *Caltha* and *Astelia* and they are intermixed frequently within the alpine community. *Lycopodium confertum* and *L. magellanicum* only seem to be present in the highest areas of the Magellanic Moorland on the island.

The maximum development of the moorland is seen on the low and gentle slopes although locally they extend to the floors of the valleys as is the case in Crossley Bay and on the tops of the hills, for example above Puerto Cook.

Fig. 3.3 Magellanic Moorland, *Astelia pumila* association (Peat bog dominated by *Astelia pumila*) Lovisato Lake (Photo: J. F. Ponce 2005)

3.1.5 Meadows Formation

The *Marsippospermum grandiflorum* association extends through valleys and places where the dense growth of rushes gives the appearance of a meadow of grasses. These meadows can be exclusively composed of *Marsippospermum* sp. or of various bushes such as *Chiliotrichum diffusum*, *Pernettya mucronata* and *Berberis ilicifolia*. *Nothofagus betuloides* and *Drimys winteri* are common but grow in stunted, bushy thickets. *Hymenophyllum dentatum*, *Asplenium dareoides*, *Galium antarcticum* and *Sphagnum* sp. are also present around the canes of the rushes. In the damp open spaces the presence of *Gunnera magellanica*, *Ranunculus biternatus* and *Senecio acanthifolius* is observed.

The most extensive development of this formation is seen in the wide valleys of Crossley Bay in the far northeast of the island (Fig. 3.4). Although it is basically a lowland formation it can extend even into the sub-alpine zone and to the hilltops, particularly throughout the open boggy areas that are wet and sheltered.

3.1.6 Alpine Formation

The alpine areas are generally located around 450 m elevation with a lower limit locally modified by edaphic factors and factors of exposure. The peaks and hilltops of the ridges above the tree line have scarce and dispersed vegetation which develops mostly in the refuges of hollows and small peat bogs, and between rocks and blocks typical in the open alpine areas (Fig. 3.5). The characteristic vegetation consists of caespitose or creeping plants. *Nothofagus antarctica* and *Empetrum rubrum* are common in this formation, particularly in the subalpine zone, but they

Fig. 3.4 Peat bog with gramineae belonging to the Meadow Formation, Crossley Bay (Photo: J. F. Ponce 2005)

Fig. 3.5 Alpine formation (Photo: J. F. Ponce 2005)

have stunted growth. Cushions and clumps of typical species of the alpine zone such as *Azorella selago*, *A. lycopodioides*, *Bolax gummifera*, *Abrotanella emarginata*, *Oreobolus obtusangulus*, *Gaimardia australis* and *Drapetes muscosoides* are seen. The creepers like *Pernettya pumila* and *Gaultheria antarctica* and the gramineae *Poa darwiniana* are also present. The small ferns *Hymenophyllum falklandicum* and *Serpyllopsis caespitose* are frequently found between rocks. *Senecio eightsii*, which is common in lower levels, appears only locally in sheltered depressions. *Drosera uniflora* appears in large marshes that are open and exposed.

3.1.7 Littoral Formation

This association develops along the rocky coasts in small peat bogs and in ground above the high water mark and in cracks in rocks in the intertidal zone. These species make up clumps or cushions that include *Colobanthus subulatus, C. quitensis, Plantago barbata, Crassula moschata, Poa darwiniana* and *Azorella filamentosa*.

Above the rocky coastline *Cotula scariosa, Ranunculus biternatus, Gunnera magellanica, Apium australe, Cardamine glacialis, Hierochloë redolens, Senecio eightsii* and *S. websteri* are common.

Some of the plants present on the cliffs are *Colobanthus subulatus, C. quitensis, Azorella filamentosa, Crassula moschata* and *Armeria maritima* which form clumps or cushions. These plants are very selective in their choice of habitat and appear sporadically on sheltered, constantly wet cliffs. Bryophytes are also very common in these places.

In the sheltered parts of the bays, between the high tide line and the forest there is a narrow and dense zone of bushes dominated by *Escallonia serrate*, occasionally accompanied by *Hebe elliptica, Berberis buxifolia* and *Ribes magellanica*. The herbaceous stratum contains *Chrysosplenium macranthum, Cotula scariosa, Gunnera magellanica, Cardamine glacialis* and *Apium australe*.

On the sandy beaches, which are not common on the island, the most common supralittoral plants are *Senecio cadidans, Caltha sagittata, Acaena magellanica, Apium australe, Hierochloë redolens, Juncus scheuchzerioides, Cardamine glacialis* and occasionally *Poa flabellata*.

3.1.8 Maritime Tussock Formation

3.1.8.1 *Poa flabellate* Association

The exposed promontories and the high ocean-facing crags are characterized by a dense vegetal association dominated by thatches of *Poa flabellate*. The thatch can be up to 2–3 m in height and has a base in the form of a fibrous mound 1–1.5 m thick, which towards the top produces a thick covering of leaves some 3 m in diameter. Occasionally *Cardamine glacialis, Apium australe* and *Blechnum pennamarina* are found. *Senecio smithii* can be abundant, particularly in the lower and wetter places. Groupings of undergrowth such as *Pernettya mucronata, Drimys winteri* and *Chiliotrichum diffusum* are scattered within this association.

3.1.8.2 *Nothofagus betuloides—Marsippospermum grandiflorum* Association

Nothofagus betuloides is the larger component, appearing as bushes and forming dense almost impenetrable thickets. The rush *Marsippospermum grandiflorum*

Table. 3.1 Correlation between the different vegetal communities recognized by Dudley and Crow (1983) and Kühnemann (1976)

Dudley and Crow (1983)	Kühnemann (1976)
Evergreen forest formation	Evergreen forest
Scrub formation	Small Gallery Forest
Magellanic moorland formation	Tundra mires Peaty grassland
Meadow formation	Grassland
Alpine formation	————
Littoral formation	————
Maritime tussock formation	————
————	Park
————	Rocky promontory

dominates the ground cover. Small poorly developed plants of *Chiliotrichum diffusum*, *Pernettya mucronata*, *Berberis ilicifolia* and *Drimys winteri* occasionally appear. The fern *Hymenophyllum tortuosum* can be found in the shelter of the twisted *Nothofagus* bushes. This vegetal association is often seen on the high slopes above the Evergreen Forest and penetrates the Deciduous Thicket of *Nothofagus* found in the elevated zones or in the Magellanic Moorland Formation of the highlands. The greatest development of this association is found on the slopes of the low round hills of Crossley Bay in the far northwest of the island.

Kühnemann (1976) as well as the seven formations of vegetation mentioned, also includes the following units of vegetation (Table 3.1).

3.1.9 Wooded Passages

Hygrophilic forests of *Nothofagus betuloides* of medium size that develop in gullies with more or less temporary water courses (Fig. 3.6). The walls of these gullies are covered with extensive networks of hepaticas. In places without trees the hepaticas and mosses are joined by gramineae such as *Poa darwinia*na and *Phleum* sp.

3.1.10 Parkland

This type of vegetation appears at the base of open valleys. It is dominated by "islands of trees" of *Nothofagus betuloides* surrounded by "Boggy meadows" and "Grassy Meadows", which will be described later. Some hardly developed trees can be seen in the zones with pools of water, and then larger groups appear forming the "islands of trees" that are oval in shape with the long axis conforming to the predominant wind direction or are circular. In the peripheral areas the trees have more foliage compared to the interior of the islets where the majority of the trunks are bare and dry. This is a xerophilic environment where mosses of all kinds appear.

Fig. 3.6   a. Wooded passages. **b**. *Nothofagus betuloides* forest. San Juan del Salvamento (Photo: J. F. Ponce 2005)

3.1.11   Rocky Promontories

These are outcrops of rocks that make up the hillsides exposed to the strong winds. In these environments the ecological conditions are extremely hard and consequently there is almost zero major vegetation. Only lichens appear on the higher parts, above the limit of the stunted forest.

In the cracks some gramineae grow as well as *Berberis buxifolia*. In more exposed places cushions formed by *Bolax gumífera* and *Nassauvia pygmaea* are observed along with patches of *Drapetes muscosus*. In the ravines some trees, stunted by the wind action, may be seen.

References

Cabrera A (1976) Regiones fitogeográficas argentinas. In: ACME (ed) Enciclopedia Argentina de Agronomía y Jardinería, vol 2, 1st edn. Buenos Aires, pp 1–85

Dudley TR, Crow GE (1983) A contribution to the Flora and Vegetation of Isla de los Estados (Staaten Island), Tierra del Fuego, Argentina. Antarct. Res. Ser. 37:1–26, American Geophysical Union, Washington, DC

Kühnemann O (1976) Observaciones ecológicas sobre la vegetación marina y terrestre de la Isla de los Estados (Tierra del Fuego, Argentina). Ecosur, Argentina 3(6):121–248

Niekisch M, Schiavini A (1998) Desarrollo y conservación de la Isla de los Estados. Unpublished report. CADIC, Ushuaia. p 70

Young SB (1973) Subantarctic rain forest of Magellanic Chile: distribution, composition, and age growth rate studies of common forest trees. Antarct Res Ser 20:307–322

Chapter 4
Geology

Abstract The oldest geological unit is the Lemaire Formation which is a stratified volcanic complex active during the upper Jurassic. This formation is composed by tuffs and acidic lavas interlayered with tuffite, tuffaceous sandstones, silty clay-stones, black mudstones and fine conglomerates. Above this formation is developed the Beauvoir Formation, formed by mudstones, siltstones, greywackes, massive quartzose sandstones and limestones. This formation represent a process of marine sedimentation would have taken place during the first half of the Cretaceous period. Post-Cretaceous sedimentary rocks are arranged horizontally to sub-horizontally over the folded strata of the Beauvoir Formation via an angular unconformity. The deposits of the Late Cenozoic are composed of till accumulations of Pleistocene age and Holocene silts, clays, sands, and gravels. The principal structure of the island is the folding of both formations (Lemaire and Beauvoir) in a large syncline, the axis of which describes a large S oriented generally east–west. The age of the folding is assigned to the middle part of the Cretaceous.

Keywords Isla de los Estados • Geology • Scotia tectonic plate • Lemaire formation • Beauvoir formation • Magellan-Fagnano system of faults

4.1 Introduction

Isla de los Estados is placed at the northern edge of the Scotia tectonic plate. It is located approximately 30 km south of the Magellan-Fagnano system of faults that define the strike-slip boundary between the South American tectonic plate to the north and the Scotia tectonic plate to the south. This plate boundary produces a lateral displacement in the order of 5 mm per year with a relative movement of the northern part of the Scotia plate towards the east (Del Cogliano et al. 2000). Towards the

J. F. Ponce and M. Fernández, *Climatic and Environmental History of Isla de los Estados,*
Argentina, SpringerBriefs in Earth System Sciences, DOI: 10.1007/978-94-007-4363-2_4,
© The Author(s) 2014

south, Isla de los Estados is bounded by a deep drop. This geomorphological feature developed as a consequence of an existing subduction regime related to the uplift of the southern end of the Andes mountain chain. This compressive regime was active during the Cretaceous—Palaeogene.

Geological studies carried out on Isla de los Estados are, up to the present day, very scarce. The first of these was performed by Harrington (1943), who distinguished the principal lithostratigraphic units that outcrop on the island. Of these the oldest he called "Porphyrica Series" and the most modern he called "Pizarreña Series", both being assigned to the Upper Jurassic. Dalziel et al. (1974a) modified some of the stratigraphic concepts of Harrington and give a different interpretation of the structures. Years later, Caminos and Nullo (1979) presented a comprehensive stratigraphy of the island.

Other works (Dalziel and Elliot 1973; Dalziel et al. 1974b, 1975) are related to Isla de los Estados in respect of its geotectonic position within the Scotia Arc.

4.2 Stratigraphy

The oldest geological unit is the Lemaire Formation which is a stratified volcanic complex formed by tuffs and acidic lavas interlayered with tuffite, tuffaceous sandstones, silty claystones, black mudstones and fine conglomerates. Together they reach a thickness of 10,000 m, showing regional metamorphism. They are folded in a large syncline with an s-shaped axis and a generally E–W orientation (Caminos and Nullo 1979).

Dalziel et al. (1974a) identifed three lithological units for this formation: type 1, tuffs and non-stratified volcanic breccias; type 2, tuffaceous sedimentary rocks and well stratified tuffs interlayered with lavas; and type 3, rocks with marine sedimentary lithology and abundant pyroclastic components.

The outcrops of this formation compose the chain of hills on the island and also the coasts of the southern edge (Caminos and Nullo 1979; Figs. 4.1 and 4.2). Its base is not exposed. The upper contact with the Beauvoir Formation is transitional and crops out only in the southwest sector of Flinders Bay (Dalziel et al. 1974a).

Caminos and Nullo (1979) assigned an age of Middle to Upper Jurassic to the Lemaire Formation, which correlates with the Patagonian-Fueguian Mountain Range to which it belongs.

The rocks of the Beauvoir Formation outcrop in the northeast sector of Isla de los Estados, in the Año Nuevo islands and on the east coast of Isla Grande de Tierra del Fuego (Dalziel et al. 1974a; Figs. 4.1 and 4.3).

This formation is composed of mudstones and slaty siltstones, greyish black in colour, greywackes of the same colour, massive quartzose sandstones, calcareous siltstones, dark grey micritic limestones and coaly, pyritiferous, blue–black siltstones. In some layers carbonaceous concretions, septarians and the remains of belemnites, pelecypods and bryozoans are found. Together they display a high level of diagenesis and incipient dynamic metamorphism, with the development

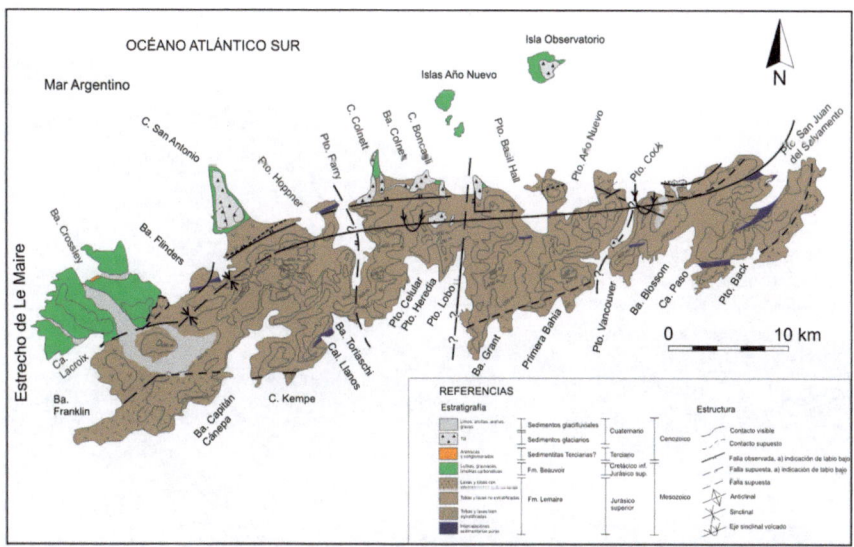

Fig. 4.1 Geological map of Isla de los Estados. (Modified from Ponce and Rabassa 2012)

Fig. 4.2 Outcrop of rocks belonging to the Lemaire Formation (Photo: Ljung 2005)

Fig. 4.3 Mudstones belonging to the Beauvoir Formation in Crossley Bay (Photo: Ljung 2005)

of slaty cleavage very well marked in some places (Caminos and Nullo 1979). Several quartz veins cross the rocks of this formation both concordantly and discordantly with the stratification (Dalziel et al. 1974a). Regarding the age of this formation, Harrington (1943) established a possible supra-Jurassic age.

Dalziel et al. (1974a) assign to this unit an Upper Jurassic-Lower Cretaceous age, tentatively correlating it with the Zapata Formation (Katz 1963), which outcrops in Chile and equivalent to the Río Mayer and Beauvoir formations of Argentina.

Caminos and Nullo (1979) inferred a maximum age of Lower Cretaceous (Valanginian-Hauterivian), correlating it with the Pampa Argentina and Rincón formations.

On the coast of Crossley Bay (54° 48′ 16″S and 64° 41′ 03.8″W), in the northeastern part of Isla de los Estados, sedimentary rocks outcrop that have been tentatively assigned by Ponce and Martínez (2007) to the Palaeogene. The deposits are made up of conglomerates and sandstones that are arranged horizontally to subhorizontally over the folded strata of the Beauvoir Formation (Upper Jurassic—Lower Cretaceous) via an angular unconformity (Ponce and Martínez 2007).

The deposits of the Late Cenozoic are composed of till accumulations of Pleistocene age and Holocene silts, clays, sands and gravels. These till deposits are found covering the base of the glacial troughs of the mountainous areas of the island and the foot of the northern slopes, where they can reach thicknesses of 15–20 m. These deposits also cover the lower zones of the western portion of the island, large parts of the capes San Antonio and Colnett and the Año Nuevo islands. Extensive deposits of till lying on the volcanic rocks of the Lemaire Formation are seen in Cape Colnett and Cape Boncagli (Colnett Bay), where the blankets of till cut by the cliff reach 50 m in thickness (Fig. 4.4). There are also concentrations of volcanic blocks (Lemaire formation) and slates (Beauvoir Formation) in this bay, of up to 2 m in diameter (Fig. 4.5). On the west coast of Cape Colnett, a moraine of 10–12 m in height with cliffs is observed (54° 43′ 51.6″S; 64° 20′ 50.0″W). The till that forms it is blue-grey with a high proportion of fine material and blocks, several of these blocks are distributed across the current beach reaching 4–5 m in diameter. The lithology of these blocks is that of andesites, porphyries, greenish silicified tuffs, greenish sandstones and black slates. In the easternmost part of Colnett Bay (54° 44′ 10.2″S; 64° 17′ 46.0″W) a notable concentration of erratic blocks is seen. These blocks have an average diameter of 80 cm and maximum diameters of up to 5 m.

In the cliffs of the western coast of Observatorio Island a layer of brecciated till of 2 m thickness is seen. It has a high level of consolidation and is deposited on the slates of the Beauvoir Formation.

Fig. 4.4 Deposits of till on the west coast of Cape Colnett (Photo: Rabassa 2005)

Fig. 4.5 Blocks of up to 2 m diameter on the beach of Colnett Bay, Isla de los Estados (Photo: Ponce 2005)

The lithology of the clasts that make up this tillite indicate a local origin with a high proportion of slates from the Beauvoir Formation and volcanics from the Lemaire Formation.

In Puerto Cook a lateral moraine is developed on the eastern sector of the fjord at some 250 m a.s.l. forming the isthmus that separates this port from Puerto Vancouver.

In Puerto San Juan del Salvamento, on the west coast of the fjord deposits of till up to 10 m thick are present, composed of larges sub-angular blocks with clayey silt matrix of bluish grey colour.

On the cliffs of the southern sector of Crossley Bay colluvial deposits are seen formed by local material and deposits of till in direct contact with the black mudstones of the Beauvoir Formation. Aeolian deposits forming longitudinal dunes are observed around the coasts of Crossley Bay (Fig. 4.6). This kind of landscape was also observed in Lacroix Cove, Franklin Bay (Ponce et al. 2011a), displaying similar characteristics. The dunes show horizons of reddish colour, rusty, associated with old positions of the phreatic level and black layers, with organic material, possibly related to old stabilization surfaces of the dunes.

4.3 Structure

The principal structure of the island is the folding of both formations (Lemaire and Beauvoir) in a large syncline (Fig. 4.1), the axis of which describes a large S oriented generally east–west (Caminos and Nullo 1979). The intensity of the folding

Fig. 4.6   A ravine cut in the dunes on the coast of Crossley Bay (Photo: Rabassa 2005)

increases from east to west. In the eastern sector of the island the fold is asymmetric and tight and is turned over towards the south in such a way that the layers of the south-eastern limb appear in an inverted position. In the central portion the fold is gentler and more open, somewhat asymmetric, reducing its intensity towards the west. In this sector the fold is also turned over towards the south with a high angle. At the western end the folding disappears and the layers are found to be sub-horizontal (Dalziel et al. 1974a, b).

Numerous secondary, minor folds are contained within the major structure. The direction of the central mountain chain and the lengthening of the island itself are parallel to the orientation of the folded structure (Caminos and Nullo 1979).

The age of the folding is assigned to the middle part of the Cretaceous, between the Upper Albian and the Coniacian, according to Caminos and Nullo (1979).

The faults found by different authors are not related to the folding structure. These faults are direct and inverse, with high angle, probably younger than the folding (Caminos and Nullo 1979).

There are sets of transverse faults of directions northeast-southwest and northwest-southeast respectively, oblique to the axes of folding. Several longitudinal faults run along the northern side of the folding. Likewise, Caminos and Nullo (1979) considered Franklin Bay to be a tectonic depression bounded to the north and south by longitudinal faults.

4.4 Geological History

The oldest geological event registered on Isla de los Estados is volcanic in character and was produced in the middle and upper parts of the Jurassic period (Caminos and Nullo 1979). This volcanic event consisted of the emission of great volumes of tuffs, acidic, calc-alkaline, rhyolitic and rhyodacitic lavas that accumulated in thicknesses several thousands of metres. During this period the landscape consisted of small islands and volcanoes that emitted large volumes of lava and ash. In the lower surrounding ground and the shallow seas that surrounded the volcanic islands the material emitted by the volcanoes and the material generated from the erosion of the emerging lands accumulated. The eruptions were explosive and, in some cases, possibly submarine. The rocks produced by these volcanic events make up the line of hills on the island and the coasts of the southern side. This group of rocks comprise the Lemaire Formation.

At the end of the Jurassic period the volcanic activity was replaced by a marine environment in which great thicknesses of sediments were produced. This process of marine sedimentation would have taken place during the first half of the Cretaceous period (Caminos and Nullo 1979). There is no evidence that between one episode and the other there was any significant crustal deformation, although it is possible that local erosive hiatuses may have existed throughout the time of contact between the two formations. The marine sediments consisted of silts, clays and carbonates. The environment in which they accumulated would have been one of calm, poorly oxygenated waters. The collection of rocks that resulted from this accumulation of sediments of marine origin is called the Beauvoir Formation and they are found in the northwest part of Isla de los Estados and in the Año Nuevo Islands.

Towards the middle part of the Cretaceous both geological formations, Lemaire and Beauvoir, were deformed into a large asymmetrical syncline verging towards the Atlantic side of the orogeny (Caminos and Nullo 1979). The current shape of Isla de los Estados responds to the geometry of this large fold elongated in an east–west direction and with a slight S shape. The outcrops of Isla de los Estados represent the edge of the mobile strip folded and turned over towards the foreland or cratonic areas, comparatively more stable, region by the extra-Andean basin or Springhill platform. The fold was the product of one single tectonic event, divided into three successive phases of deformation and accompanied by low grade dynamic metamorphism. This phenomenon transformed the volcanic rocks and associated sediments into slates, phyllites and shale tuffs.

Just as with the Fuegian mountains, Isla de los Estados suffered an intercontinental bowing of the mobile strip, which modified its original orientation from N–S to E–W, in order to form the northern wing of the Scotia Arc. According to Dalziel and Elliot (1973) this occurrence took place around the Cretaceous–Tertiary boundary.

During the Palaeogene period the dominant environment of Isla de los Estados would have been continental with the presence of rivers and torrential streams at the foot of a line of mountains, which would have been growing as a consequence of the interaction of the tectonic plates that continued up to the end of this geological period. A small part of the result of this new process of accumulation of fluvial sediments is the collection of sedimentary rocks found in Crossley Bay in the north-western end of the island.

Already towards the Pleistocene period Isla de los Estados constituted a collection of highlands that marked the south-eastern end of the Fuegian Andes.

The deposits of sediments and rocks accumulated during the last few million years on Isla de los Estados, just as on Isla Grande de Tierra del Fuego, are principally the result of glacial action. In the passing of the last 7 million years, Isla de los Estados, as with a large part of the planet, has been submitted to the action of several glacial events, the last of which ended approximately 18,000 years ago. This last glacial event, which reached its maximum development approximately 24,000 years ago (Last Glacial Maximum or LGM), generated the majority of the sediments of glacial origin that can be seen today throughout Isla de los Estados. The accumulations of till are found covering the floors of the large glacial valleys in the mountainous area and at the foot of the slopes of the northern side of Isla de los Estados, where they reach thicknesses of 15–20 m. These deposits also cover the low lying areas of the western sector of the island and most of capes San Antonio and Colnett and the Año Nuevo islands.

References

Caminos R, Nullo F (1979) Descripción Geológica de la Hoja 67 e, Isla de los Estados. Territorio Nacional de Tierra del Fuego, Antártida e Islas del Atlántico Sur. Servicio Geológico Nacional. Boletín 175. Buenos Aires, p 52

Dalziel IWD, Elliot DH (1973) Scotia arc and antarctic margin. In: Nair AEM, Stehil FG (eds) Ocean basins and margins, vol 1, New York, pp 171–245

Dalziel IWD, Caminos R, Palmer KF, Nullo FE, Casanova R (1974a) South extremity of Andes: Geology of Isla de los Estados, Argentina, Tierra del Fuego. Am Assoc Petrol Geol Bull 58(12):2502–2512

Dalziel IWD, de Wit MJ, Palmer KF (1974b) Fossil marginal basin in southern Andes. Nature 250:291–294

Dalziel IWD, Dott RH, Winn RD, Bruhn RL (1975) Tectonic relations of South Georgia Island to the southernmost Andes. Geol Soc Am Bull 86:1034–1040

Del Cogliano D, Perdomo R, Hormaechea JL (2000) Desplazamiento entre placas tectónicas en Tierra del Fuego. XX Reunión Científica de la Asociación Argentina de Geofísicos y Geodesias, Actas, Mendoza, pp 231–234

Harrington HJ (1943) Observaciones geológicas en la Isla de los Estados. Anales Museo Argentino de Ciencias Naturales. Geología 29:29–52. Buenos Aires

Katz HR (1963) Revision of cretaceous stratigraphy in patagonia Cordillera of Ultima Esperanza, Magallanes Province Chile. AAPG Bull 47:506–524

Ponce JF, Martínez O (2007) Hallazgo de depósitos sedimentarios postcretácicos en Bahía Crossley, Isla de los Estados, Tierra del Fuego. Revista de la Asociación Geológica Argentina 62(3):467–470

Ponce JF, Borromei AM, Rabassa JO, Martínez O (2011a) Late quaternary palaeoenvironmental change in western Staaten island (54.5°S, 64°W) Fuegian Archipelago. Quatern Int 233:89–100

Ponce JF, Rabassa J (2012) Geomorfología glacial de la Isla de los Estados, Tierra del Fuego, Argentina. Revista de la Sociedad Geológica de España 25(1–2):67–84

Chapter 5
Glacial Geomorphology

Abstract Based on a geomorphological analysis, a timeless glacial model of the island is presented, that is, a model that could be applied to any of the Pleistocene glaciations of the region. This model is basically oriented towards the formation of large valley glaciers, fed by cirque glaciers and small, local ice caps. Several of these glaciers excavated their troughs following the orientation of tectonic alignments, such as faults and folds and also, stratigraphic, intra-formational boundaries. In the southern sector of the island, the glaciers became in contact with the ocean quite soon, ending in an ablation process of the "calving" type, that is, the formation of icebergs. The glaciers of the northern coast, instead, were flowing nested in deep valleys until they reached the extensive plain located north of their mountain sources, a plain that had been covered by the sea during the interglacial periods, as it happens today. There, these ice bodies ended as piedmont glaciers, in an ablation process by simple melting of their terminal portions.

Keywords Geomorphology • Glacier • Periglacial features • Fjords • Isla de los Estados • Patagonia

5.1 Introduction

During the Last Glacial Maximum (LGM; ca. 24 cal. ka B.P.; Rabassa 2008), a large part of Patagonia was covered by glaciers. An enormous blanket of ice formed in the Andes mountain range from the north of the province of Neuquén to the southern end of Tierra del Fuego (Fig. 5.1a), covering a surface of approximately 320,000 km^2, that is to say, a size equivalent to one-third the current surface area of Argentine Patagonia (Ponce and Rabassa 2012b). Towards the west, south of parallel 42°S, the ice covered to whole of the surface that today we know as the Republic of Chile. From the mountain ranges, huge tongues of ice descended towards the Pacific Ocean. In some sectors, the ice came into direct contact with the sea possibly generating small platforms of ice attached to the

J. F. Ponce and M. Fernández, *Climatic and Environmental History of Isla de los Estados,* 45
Argentina, SpringerBriefs in Earth System Sciences, DOI: 10.1007/978-94-007-4363-2_5,

Fig. 5.1 a Patagonia during the LGM with sea level −120 m. **b** Extension of the Beagle Paleoglacier during the LGM

coast and a great number of enormous icebergs (Ponce and Rabassa 2012b). Large volumes of fresh water were produced along the entire Pacific Ocean coast, coming from the melting of these glaciers, which must have seriously affected the coastal marine ecosystems in these times. Towards the sector east of the Andes, the ancient Patagonian ice sheet unloaded large volumes of ice through extensive glaciers whose fronts reached the Andean foothills. Several of these old glaciers occupied the surface of lakes currently located at the feet of the Andes (for example, the lakes Fagnano, Argentino, Viedma, Buenos Aires, Nahuel Huapi, Lácar, among others). Throughout the Argentine provinces of Santa Cruz and Tierra del Fuego, some glaciers exceeded 100 km in length in a west–east direction.

5.2 Beagle Channel, Isla Grande de Tierra del Fuego

The Beagle Channel (54° 53′S; between 66° 30′ and 70°W), some 120 km to the west of Isla de los Estados, is located to the south of Isla Grande de Tierra del Fuego (Fig. 5.1a), and constitutes a tectonic valley that was completely filled with ice during the Last Glacial Maximum. The so-called "Beagle Glacier" came from the Mountain Ice Field of the Darwin Mountain Range (Tierra del Fuego, Chile), receiving tributary glaciers all along its length from the cirques and valleys in the interior of the lines of mountains found on either side. During its greatest extent, the front of the ice would have been located close to Punta Moat (Fig. 5.1b), at the eastern end of the Beagle Channel (approximately 120 km to the west of Isla de los Estados). Judging by the radiocarbon dates of the basal peat bogs along the Beagle Channel, it can be inferred that around 15 ^{14}C ka B.P. (ca. 17 cal. ka B.P.) the ice front had retreated some 100 km towards the west of its point of maximum extension. The definitive retreat of the ice was produced at least around 10 ^{14}C ka B.P., when the first plant communities established themselves in the steppe/tundra environment (Markgraf 1993; Heusser 1989a, b).

5.3 Geomorphology of Isla de los Estados

The first geomorphological references concerning Isla de los Estados are the studies carried out by Caminos and Nullo (1979) and García (1986). In both studies the principal geomorphological features of the island are described, dividing it into two main regions: one mountainous, central and eastern and the other western and with flatter morphology.

According to Caminos and Nullo (1979), to the west of the Spegazzini hills the landscape has landforms of gentle relief and low hills (Fig.5.5). In this sector the outcropping rocks are principally those of the Lemaire and Beauvoir formations. The relief is made up of rounded and flat-topped forms, the valleys are wide with gently sloping sides, they have a maximum width of 3.3 km. The floor is flat, filled with fluvio-glacial deposits (Caminos and Nullo 1979), and a large percentage of this floor is currently occupied by extensive peat bogs. The maximum heights do not exceed 500 m a.s.l. and these high points increase from west to east.

In Crossley Bay basal till deposits can also be seen, generating a flat relief.

The central-eastern region of the island contains a large variety of landforms, products of glacial action (Fig. 5.6). They comprise a glacial landscape of Alpine type. This landscape was generated on rocks of the Lemaire Formation. Caminos and Nullo (1979) describe a great number of cirques and troughs for this sector, as well as jagged crests, needles, ridges and pinnacles.

5.3.1 Periglacial Features on Isla de los Estados

The areas dominated by periglacial cold environments generate characteristic structures. Their study in active environments reveals what the conditions are in which they develop. The existing knowledge on these cryogenic landforms located above the tree line in the Fuegian Andes is very limited and generalised. Auer (1970), referring to the area of Ushuaia (54° 50'S), cited the occurrence of landforms related to seasonal freezing and the penetration of freezing to a maximum depth close to 0.3 m at elevations around sea level. On Monte Louis Martial, close to the city of Ushuaia, he suggested that the lower limit for patterned ground of small scale is found at 750 m a.s.l., whereas for the large polygonal networks, it is located above 900 m a.s.l. In the central portion of the Fuegian Andes, Garleff (1977) located the lower limit of the solifluction and patterned ground close to 700 m a.s.l. Corte (1977) included Tierra del Fuego in the Argentine Andean region where the lower limit of the permafrost was estimated at 900 m a.s.l., using a combination of observations of freeze–thaw structures of 1 m diameter and an estimation of the average annual air isotherm of 0 °C close to 950 m a.s.l. Recently, Valcárcel-Díaz et al. (2006) described several types of periglacial landforms in the area of Monte Alvear, in the the Fuegian Andes. All these landforms were found in a range of altitudes that varied between 1000 and 1050 m a.s.l.

During a Swedish/Argentine field campaign, carried out in December 2005, for the first time stone stripes and sorted circles were found on Isla de los Estados (Ljung and Ponce 2006). These periglacial features were observed above the present tree line (400 m a.s.l.) at approximately 500 m a.s.l., in San Juan del Salvamento at the eastern end of the island.

The stone stripes (54° 45' 47.6"S; 63° 53' 41.3"W) have a thickness close to 15 cm, the sediment that they are composed of is fine grained in the central part, surrounded by depressions filled with coarser material (Fig. 5.2). They are oriented parallel to the slope of the hillside where they are formed. The sorted circles have an average diameter of 10 cm with fine grained sediment in the centre surrounded by depressions with coarser grained material (Fig. 5.3).

The periglacial features observed on Isla de los Estados were produced by repeated cycles of freezing–thawing. These claims are based on the presence of a central sector composed of fine grained sediments and edges with a coarser grain size and the fact that the sorted circles are found in groups with the central part elevated. These characteristics are commonly encountered in features formed by freezing action (Goldthwait 1976). The lack of vegetal cover on these periglacial features due to a constant removal indicates that they are active cryogenic features. The covering of snow during the winter months is probably limited as a

Fig. 5.2   Stone stripes at 500 m a.s.l. in San Juan del Salvamento (Photo: Ljung 2005)

Fig. 5.3   Sorted circles at 500 m a.s.l. in San Juan del Salvamento (Photo: Ljung 2005)

consequence of being an exposed area. This is similar to that which is observed in other Sub-Antarctic islands where the development of periglacial features is also associated with exposed areas with little snow cover and a limited isolation of the air during the coldest months (Belhouvers et al. 2003).

5.3.2 Glacial Morphology

Recently Ponce et al. (2009) and Ponce and Rabassa (2012a) carried out a characterisation of the principal erosive glacial landforms present on Isla de los Estados through a morphometric analysis of these landforms, constructing a geomorphological map (Fig. 5.4) and a non-time constrained glacial model. There now follows a transcription of the most important aspect of both these works.

5.3.2.1 Cirques

Cirques are semi-circular depressions in the shape of amphitheatres associated with the modelling of the heads of the glacial valleys (Figs. 5.5, 5.6 and 5.7). They constitute the source of Alpine type glaciers.

The expansion and recession of the glaciers during the Late Cenozoic occurred in response to the vertical fluctuations of what is called, in a broad sense, the snow line, produced by global and/or regional climate changes (Porter 1975).

The permanent snow line, to differentiate it from the "transitory snow line", is a line or altitude of land that separates areas where fallen snow disappears in summer from areas where remnants of the snow remain all year round (Martínez 2002).

The study of cirques and populations of cirques usually provides valuable glaciological and climatic information and usually consists of establishing three types of attributes: shape of the cirque, size and exposure (Haynes 1995). Obtaining the mean altitude of the bases of the fossil cirques is a good minimum indicator of the position

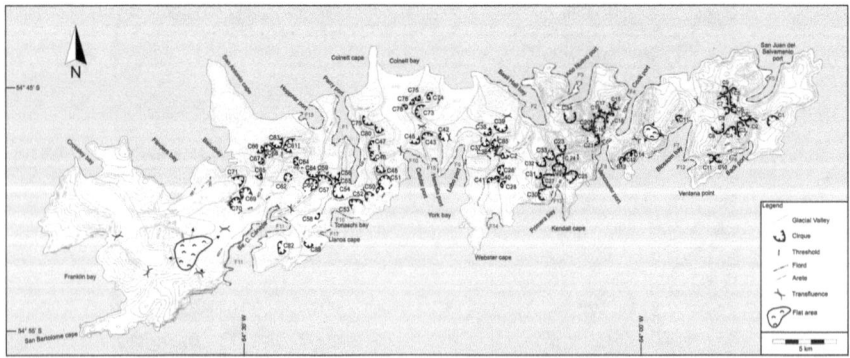

Fig. 5.4 Geomorphological map of Isla de los Estados (modified from Ponce and Rabassa 2012a)

Fig. 5.5 Cirques in San Juan del Salvamento (Photo: Ponce 2005)

Fig. 5.6 Cirque in Flinders Bay (Photo: Ponce 2010)

of the old snow line (Flint 1957; Embleton and King 1968). This method has been widely used (Charlesworth 1957; Flint and Fidalgo 1964; Andrews 1975; Meierding 1982) in various regions of the world. It is agreed that the maximum altitude of the permanent snow line can be deduced when it is certain that the glacier extended beyond the cirque, and consequently it can be assumed that the feeding zone extends beneath this position (Flint 1957).

Fig. 5.7   Cirques in Colnett Bay (Photo: Ponce 2005)

The total number of cirques present on Isla de los Estados is 79. Whilst the total number of cirques is high in relation to the surface area of the island, these landforms in most cases are not well preserved and it is difficult to distinguish all the component parts: headwall, basin, threshold (Fairbridge 1968).

The range of altitudes that the basins of the cirques have on Isla de los Estados is very wide, varying between 0 and 460 m a.s.l. These surfaces are currently found lacking in bodies of ice or snow, showing that the current equilibrium line altitude (ELA) or permanent snow line is above 800 m a.s.l. (maximum altitude on the island). Around 72 % of the cirques in the study area have heights between 200 and 400 m a.s.l. The average height for the bases of all the analysed cirques is 272 m a.s.l., placing the old permanent snow line at a height close to this average value.

The total area occupied by the cirques on Isla de los Estados is 36.5 km^2; this value represents 7 % of the total surface area of the island, which is a very high value for islands of intermediate dimensions. The dimensions of these cirques are varied, displaying an average surface area or 0.46 km^2, with a maximum of 1.29 km^2 and a minimum of 0.008 km^2. Their shapes are principally circular and semi-circular producing a relationship between the transverse and longitudinal axes (T/L) that varies between 0.5 and 2.3, with an average value of 1.2. The average diameter of the cirques is 0.71 km.

The predominant direction of exposure is towards the southeast with 23 % of the cirques oriented in this direction. The number of cirques associated with the south coast of Isla de los Estados is 38 and they have an average surface area of 0.37 km^2.

Composite cirques have been seen, that is to say, small cirques developed on the upper parts of larger cirques. This relation, called "cirque-in-cirque" is often related to changes in the position of the permanent snow line after the Last Glacial Maximum.

There is no relationship seen between the location of the cirques and the lithological substrate, due to the fact that the elevated zones of the island are almost exclusively represented by tuffs and lavas of the Lemaire Formation with only some minor lithological variations (Caminos and Nullo 1979).

5.3.2.2 Troughs

Glacial erosion is a very effective process when it acts on rocky beds. Rates of erosive effectiveness been estimated between 10 and 20 times greater than fluvial erosion processes and mean flooding between 0.005 and 3.00 mm/year (Bermúdez et al. 1992). The main erosive landforms produced by this phenomenon are the troughs, characterised by steep walls, straight routes and for their U-shaped cross-section. These forms of relief are the most distinctive and abundant feature of the landscape on Isla de los Estados (Figs. 5.8, 5.9 and 5.10) considering that the total surface area occupied by this type of landform is 220 km^2, representing 42 % of the surface area of the island. Once again, this is a very high value for this parameter, surely connected to a very significant glacial activity of the archipelago during the LGM. A total of 48 troughs have been described on the island of which 18 currently form fjords through their extension into the sea.

The dimensions of the troughs on Isla de los Estados are very varied, giving an average surface area of 4.6 km^2, with a maximum of 38 km^2 (trough 35) and a minimum of 0.5 km^2 (trough 22). The average length is 3.26 km with a maximum of 15.4 km (trough 35) and a minimum of 0.64 km (trough 22). The northern coast has troughs of greater size, possibly due to the presence of an extensive piedmont flatland, currently submerged and a gentler relief than in the southern part of the island, where

Fig. 5.8 Lake Lovisato (trough A19) (Photo: Ponce 2010)

Fig. 5.9   Hanging valley tributary to the trough of Lake Lovisato (Photo: Ponce 2005)

Fig. 5.10   Crossley Bay (trough A27) (Photo: Ponce 2010)

the majority of the troughs do not reach great extensions. Several of the larger troughs display a continuity of their erosive features for several kilometres below the current sea level. In the central portion of the island (for example in Colnett Bay), this characteristic becomes more significant, reaching values close to 5 km of submarine length. In this sector the isobaths have shapes typical of a wide shallow valley.

The troughs on Isla de los Estados are of Alpine type. Although the landforms have typical wide U-shaped cross-sections, in many cases these shapes are not clear and a significant amount of erosion is seen that occurred after their development and masks their original form.

Troughs are seen that have no associated cirque. In these cases, their sources are found associated to relatively high and flat areas (for example troughs A6 and A6′, approximately 200 m a.s.l.) that, it seems, have acted as zones of accumulation and feeding of these troughs. Their transverse profiles show a maximum width close to 2 km. They have the typical curved shape with walls that reach 200 m in height and with slopes of 13°.

The values of the angle of the slope of the troughs varies between 14° (trough 35) and 45° (West slope of Puerto Parry). The longitudinal slopes show an average value of 4.7°. The troughs associated with the south coast of the island have a greater longitudinal slope (an average value of 5.4°) than those associated with the north coast, whose average value is 4.1°.

The trough of San Juan del Salvamento developed following the route of a sedimentary wedge. The troughs of Puerto Cook (A39) and Puerto Vancouver (A40) clearly developed on a inferred fault that crosses the island from SSW to NNE. The trough of Puerto Año Nuevo (A45) follows, for over half its length, the direction of a fault that forms the contact between sedimentary rocks and lavas and tuffs of the Lemaire Formation. Trough A14 develops almost in its entirety on a N–S oriented fault. On this same fault the north running trough of Puerto Victoria developed (A17). These are some of the many examples of a strong structural control that is seen for the troughs on the island.

5.3.2.3 Fjords

Fjord is a Norwegian term universally adopted for an arm of sea of a certain size, characterised by a disposition more or less rectilinear, steep rocky sides and great depths (Fairbridge 1968) that have been the result of the flooding of a glacial trough due to the post-glacial sea level rise. Fjords are exclusively found along the marine coasts of elevated relief moulded by Pleistocene glaciers. The absolute and unmistakeable relationship of the fjords with old glaciated areas restrict their location to the high latitudes: Norway, Scotland, Greenland, Iceland, the Svalbard islands, other Arctic Ocean islands, Canada, Alaska, Kamchatka, Chilean Patagonia, New Zealand and some minor archipelagos, such as the islands of Kerguelen, South Georgia, or South Shetland (Syvitski and Shaw 1995), as well as the Antarctic Peninsula, where they have achieved a remarkable development.

The term "fjärd" makes reference to similar coastal marine incursions, but associated with glaciations in lowlands (Ahlmann 1919; Embleton and King 1968; Klemsdal 1982; Syvitski et al. 1987). The fjärds are distinguished from the fjords by their irregular shapes and lack of elevated relief and U-shaped cross-sections, all of which are characteristic of fjords (Fairbridge 1968).

The total number of fjords on the island is 18 (Fig. 5.4). The dimensions are varied, with an average length of 3.1 km, having a maximum of 7.1 km (Puerto

Parry) and a minimum of 0.75 km (Puerto Heredia). The average fjord area is 3.2 km². Puerto Basil Hall is the largest fjord with and area of 8.4 km². As in the case of the troughs, the largest fjords have developed on the northern coast of the island, possibly related to the existence of a wide current submarine platform that during the last glaciation constituted a flatter and larger emerged relief in this sector, favouring the longitudinal development of glacial valleys. The transverse profiles of these fjords reveal shapes of broad troughs with very abrupt side walls, the heights of which vary between 150 and 400 m a.s.l.

Crossley Bay and Lacroix Cove (west end of Isla de los Estados, Fig. 5.4) have the geomorphological characteristics of a fjärd. Both marine inlets are shallow, Crossley Bay having a maximum depth of 20 m and Lacroix Cove 50 m. These landforms are found in an area of relatively low relief. The height of the sides in both cases varies between 50 and 100 m a.s.l. Crossley Bay has an area of 2.8 km² with a length of 2 km and 1.8 km maximum width while Lacroix Cove has an area of 2.5 km², being 2.1 km long and 1.6 km wide. These dimensions are small in comparison with those of the fjords present on the island.

Puerto Cook (Fig. 5.11), Puerto Vancouver and Puerto Año Nuevo are clearly associated with NE-SW and N–S running faults. Puerto Lobo developed along a fault oriented N–S. The development of the trough of Puerto San Juan del Salvamento (Fig. 5.12) is related to the presence of a sedimentary wedge of the base rock, delimited by SW-NE faults. Its orientation also coincides with the axis of the main overturned anticline at the east end of the island. Despite having no faults described in Puerto Parry (Fig. 5.13) and Puerto Hoppner, the N–S direction of these fjords and their extension suggests the existence of some kind of fault, similar to that seen at Puerto Lobo, in the centre of the island. Capitán Cánepa Bay to the west, similarly to Blossom Bay (Fig. 5.14) in the eastern sector, appears to be related to the direction of the principal folding. In general, the relation between fjords and structural geology in the western sector is not as clear as in the rest of the island.

The longitudinal extension of the fjords beneath the current sea level, according to the analysis of the available bathymetric charts (Fig. 5.15) and digital models

Fig. 5.11 Puerto Cook Fjord (Photo: Ponce 2005)

Fig. 5.12 Puerto San Juan del Salvamento Fjord (Photo: Ljung 2005)

Fig. 5.13 Puerto Parry Fjord (Photo: Ponce 2010)

of height of the land, reach a maximum of 6.2 km and a depth of 180 m, such as that observed in Blossom Bay, located on the south coast of the island. The maximum depths of the bases of the fjords vary between 188 (Puerto Parry) and 11 m (Llanos Cove).

No significant variations are seen with respect to the lithology within which the fjords have developed. They consist mostly of non-stratified tuffs and massive lavas of the Lemaire Formation. Only Puerto Año Nuevo, Puerto San Juan del Salvamento and Llanos Cove show different lithologies. These three fjords follow the development of wedges formed by purely sedimentary interlayering that also

Fig. 5.14 Blossom Bay Fjord (Photo: Ponce 2005)

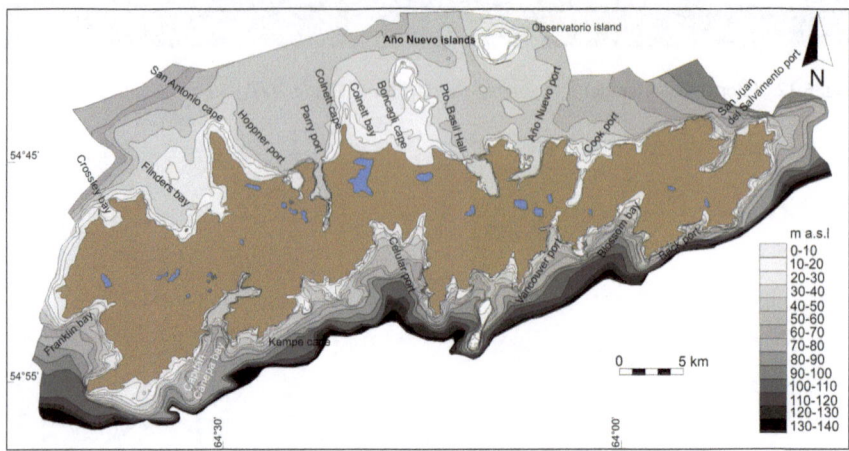

Fig. 5.15 Bathymetric map of Isla de los Estados. (modified from Ponce and Rabassa 2012a)

belong to the Lemaire Formation. The fjärds Crossley Bay and Lacroix Cove have developed almost exclusively on the mudstones of the Beauvoir Formation.

5.4 Relationship Between the Structural Geology of the Island and its Principal Geomorphological Features

The presence of two main morphological regions described by Caminos and Nullo (1979) and García (1986) is controlled by the development of a large syncline that crosses the entire island in a generally west-east orientation. The intensity of this principal folding determines the topographic characteristics of the island. The mountainous central-eastern region is associated with the zone of most intense folding. In the central zone, where the layers of the southern limb of the syncline are found in an almost vertical position (Caminos and Nullo 1979), the highest points on the island are found. Towards the eastern end, the presence of a syncline turned over towards the south limits the development of altitude in the line of mountains, reducing the maximum heights towards this sector, for which reason the layers of the southern limb slope 45° towards the south (Caminos and Nullo 1979). The western region of rolling topography is connected with a lower intensity of folding. In the western extreme of the island where the relief is gently rolling, the folding disappears and the layers are seen in horizontal or sub-horizontal position. The fact that the main syncline is turned over towards the south, imposes on the southern sector of the island a more abrupt and steeper topography than on the northern coast where the intensity of the folding reduces until it disappears in the current continental platform.

Analysing together the erosive glacial landforms described in this work, a general and strong structural control is observed in the development, location and orientation of these landforms. In the case of the troughs and fjords, the predominant orientation is SW-NE, similar to the direction of the axis of the main fold in the eastern and western sectors of the island and one of the two main sets of faults. Several of the troughs and some fjords are also associated with the presence of sedimentary wedges and contacts between different lithological types. In the particular case of the fjords, in the western portion of the island, the direction is clearly SW-NE, in the central zone the direction turns N–S and towards the east the fjords once again acquire a SW-NE orientation. In the central zone the fjords are apparently controlled by the presence of faults with a general N–S direction. Possibly this last set of faults are acting as a divider of blocks corresponding to the principal fold structure, generating small tectonic pillars.

The presence of an overturned anticline with continuity along all the longitudinal axis of the island controls in turn the sizes of the fjords and troughs to the north and south of this axis. In the southern sector, where the intensity of the folding is greater and consequently the general slope is also greater, the fjords and troughs achieve a lesser longitudinal development in comparison with those associated with the north coast, where the presence of gentler slopes adjacent to the continental platform allows a greater longitudinal development of these landforms.

The exposure of the cirques also shows a marked correlation with the existing structural geology. Almost 46 % of the cirques are oriented NW–SE, coincident with one of the two sets of transverse faults that exist on the island.

The impossibility of establishing, until now, the presence of sedimentary deposits belonging to elevated Middle Holocene marine levels along the coasts of Isla de los Estados would indicate a possible local, or maybe even regional, Late Holocene tectonic descent. The analysis of the height of the bases of the cirques shows that 26 % of these are at a height less than or equal to 200 m a.s.l. Rabassa et al. (2003) mentions the existence of differential tectonic behaviour for the southern coast of Isla Grande de Tierra del Fuego. These authors propose the presence of tectonic blocks with relative ascending and descending movements that would have taken place along the regional fractures. These fractures would have been especially active in the Late Holocene, with the genesis of levels of marine beaches elevated towards the western sector (to the west of Estancia Harberton and up to the Chilean border) and on the contrary, towards the eastern sector, the absence of elevated marine levels and presence of remainings of a submerged forest in Sloggett Bay, which is found below the current sea level, with tree trunks visible in their living positions in the intertidal zone. This situation could be repeated in Isla de los Estados, maybe with an even greater magnitude that would have favoured the flooding in great magnitude of many of the troughs with the consequent development of the fjords here described.

5.5 A Glaciation Model for Isla de los Estados

The formulation of a glaciation model for Isla de los Estados that would explain the spatial disposition of the landforms in relation to specific glacial events requires a great number of absolute dates from the identified glacigenic sedimentary deposits, something that is not currently available. However, it is possible to put forward a "non time constrained" model for Isla de los Estados, i.e. without associated chronology. From some radiocarbon and OSL dating of glacigenic sedimentary sequences on the northern coast of Isla de los Estados recently presented in Moller et al. (2010), significant glacial activity in this archipelago during the LGM was inferred. Consequently, the glacial model formulated here could be valid for the last glacial cycle (equivalent to the Wisconsin or Weichsel state, as it will be explained later) (Fig. 5.16).

In agreement with the geomorphological analysis carried out, many of the erosive glacial landforms analysed in the present work appear to be related to old glaciations, probably the Great Patagonian Glaciation (GPG; Early Pleistocene, ca. 1 Ma B.P.; see Rabassa 2008) or even earlier ones, although a large number of these landforms appear to have been reoccupied during the successively more recent Quaternary glacial events. In any case, the glaciers that moulded the island would have had a local origin, fed by a great number of existing cirques on the island and maybe by a local ice cap, unconnected to the mountain ice sheet of the Fuegian Andes. None of the cirques are currently occupied by ice, indicating that the permanent snow line is now found clearly above 800 m a.s.l. (the height of the highest point on the island), although it is possible that it might be only slightly above this elevation, given that frequent snowfalls of significant depth occur. If a lowering, even only slightly, of the permanent snow line would take place in the future, these snow patches would become glaciers in quite a short time.

From the cited accumulation zone, the lobes of ice that covered a large part of the island canalised following various main axes, and among these, the San Juan del Salvamento, Blossom Bay, Puerto Cook-Puerto Vancouver, Puerto Basil Hall, Puerto Roca, Colnett Bay, Puerto Parry, Capitán Cánepa Bay and Crossley Bay depressions. The bathymetric map of Isla de los Estados (Fig. 5.15) shows the continuity of some of these major troughs several metres below current sea level, indicating a minimum possible extension for these glacial lobes in the past.

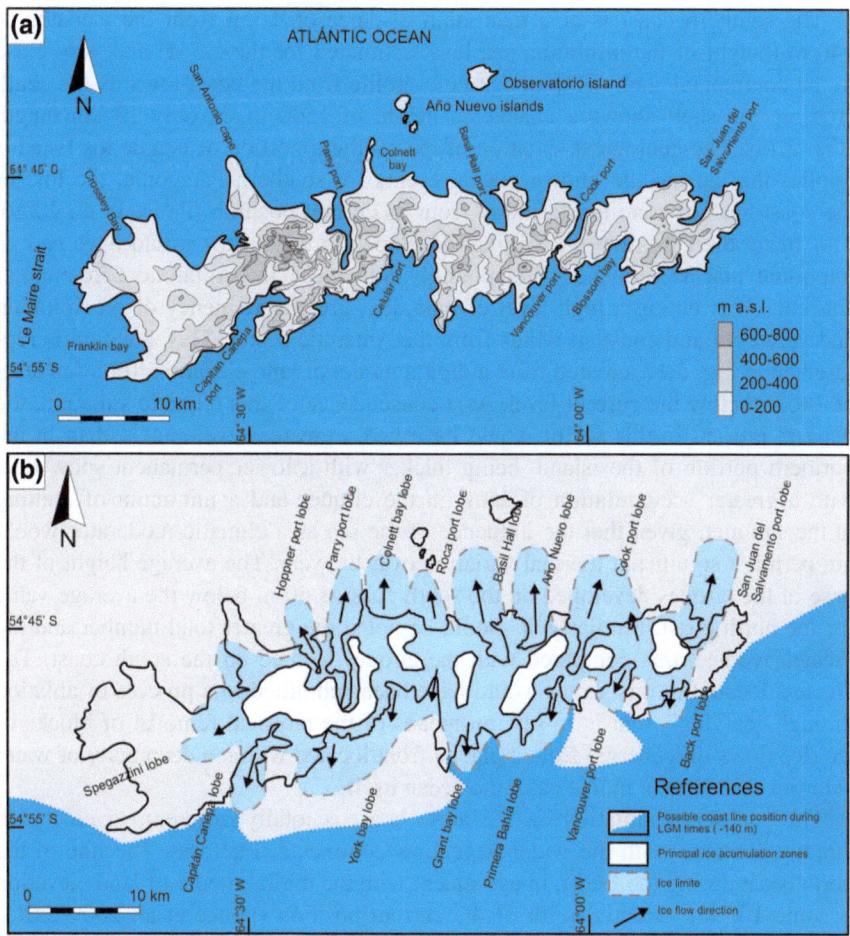

Fig. 5.16 A "non time constrained" model of glacial development for Isla de los Estados with locations of the principal glacial lobes or tongues. (modified from Ponce and Rabassa 2012a)

During the glacial and interglacial cycles significant glacioeustatic movements took place, including lower sea levels during the glacial periods, of at least several tens of metres during the cold events and more than 100–140 m during the maximum development of the subsequent glacial events (Rabassa et al. 2005).

During the Last Glacial Maximum (LGM, ca. 24 cal ka B.P.; Rabassa 2008), the sea level was positioned between 120 and 140 m below its current level (Uriarte Cantolla 2003). The configuration and position of the coasts of Isla de los Estados throughout the different glacial events would not have suffered significant variations with respect to the LGM (Fig. 5.16).

During the different glacial events differences would have existed in the dominant type of glaciation of the northern and southern coasts of Isla de los Estados.

The southern coast is at a maximum distance of 4 km from the isobaths of 140 m (height of the minimum sea level estimated for the LGM) and very close to the continental shelf. The topographic profile from the coast towards the south crossing the shelf shows a change in height of 1500 m, in only 13 km length (Fig. 5.17). The geological situation of the southern portion of Isla de los Estados implies that during the known glacial events for southern Patagonia, the line of the coast has not gone too far away from its current position (Ponce et al. 2009). This being the case, the glaciers that developed in this sector would have had an important marine influence, and possibly influenced the Antarctic Circumpolar Current, contributing a high level of humidity, greater frequency of precipitation and very cold and constant winds from the Antarctic sector. This situation is represented in Fig. 5.17, created from a digital model of land elevation, for a sea level of 140 m below the current level. As a consequence of this climatic situation, the glaciers present in this sector would have had a greater development than in the northern portion of the island, being thicker with a lower permanent snow line, with a greater accumulation of snow in the cirques and a minimum of melting in the summer, given that the influence of the sea as a climatic moderator would not permit a significant thermal variation over the year. The average height of the base of the cirques developed on the south coast is 50 m below the average value for the north coast. Similarly, it should be noted the greater total number and the greater average area compared with the cirques located on the south coast. The glaciers located in this sector would have been submitted to a process of ablation through "calving", that is to say, by means of the physical removal of blocks of ice (ice flows or icebergs) from a glacial front located within a deep body of water (Martini et al. 2001), in this case, the ocean itself.

The geological situation on the north coast is totally different, given that it directly connects with the wider Argentine Continental Platform. The line of the north coast during the LGM, in agreement with the digital model of land elevation is some 100 km directly north of the current position (Ponce et al. 2009, 2011, Fig. 5.17b).

During the different glacial events that occurred, a large part of the submarine platform would have been repeatedly exposed as a consequence of the large relative drop in sea level, forming an enormous flatland along the east coast of Pampa and Patagonia (Rabassa et al. 2005). In this way, the climate in this sector acquired greater continental environmental conditions during the cold periods. Rabassa et al. (2005) proposed a displacement of the Malvinas Current towards the east due to a reduction in the sea depth between Patagonia and the Malvinas/Falklands islands, as a consequence of lowering sea level during the glacial events. This displacement would have provoked an increase in the climatic continental environmental conditions throughout the littoral sector of the southern end of South America. Under these climatic conditions, the glaciers located on the northern coast of the island would have developed less than those in the southern sector, finishing on a nearby part of the broad flatland, in a process of ablation by simple melting of the end section. Towards the north, the majority of the glaciers may never have extended beyond 7–8 km north of the line of the current coast due to

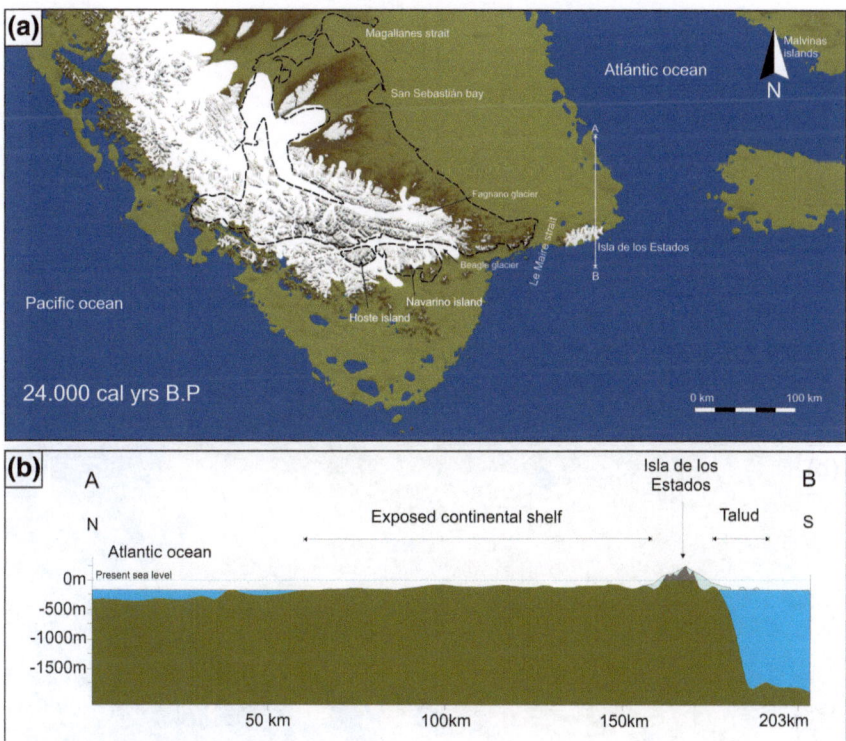

Fig. 5.17 a Extension of the ice sheet and output lobes during the LGM in Tierra del Fuego, with a hypothetic sea level 140 m below the current level. **b** Topographical profile that shows the situations of the northern and southern coasts of Isla de los Estados during the LGM. (modified from Ponce and Rabassa 2012a)

the presence of a submarine depression that extends longitudinally across from the northern coast of Observatorio Island. This depression could be associated with the tectonic sinking related to a fault zone parallel to the principal axis of folding of the island (Caminos and Nullo 1979). This strip must have been deepened by the glacial erosion produced by the bodies of ice emerging from the mountainous front when they found a large change of slope in their path. During the periods of deglaciation, this depression could have been occupied by a proglacial lake of ice waters that received the output of at least five principal glaciers. This paleolake would have had a maximum depth of 100 m and a surface area of close to 90 km^2 (Fig. 5.18). Proglacial lakes of lesser magnitude, associated with processes of glacial erosion, would have also developed at the fronts of the lobes of Puerto Parry, Puerto Hoppner and Puerto San Juan del Salvamento. Whereas no evidence has as yet been collected from the current submarine platform to support this hypothesis, the geomorphological analysis of the sea floor and the generation of digital models of elevation from bathymetric data strengthen this idea.

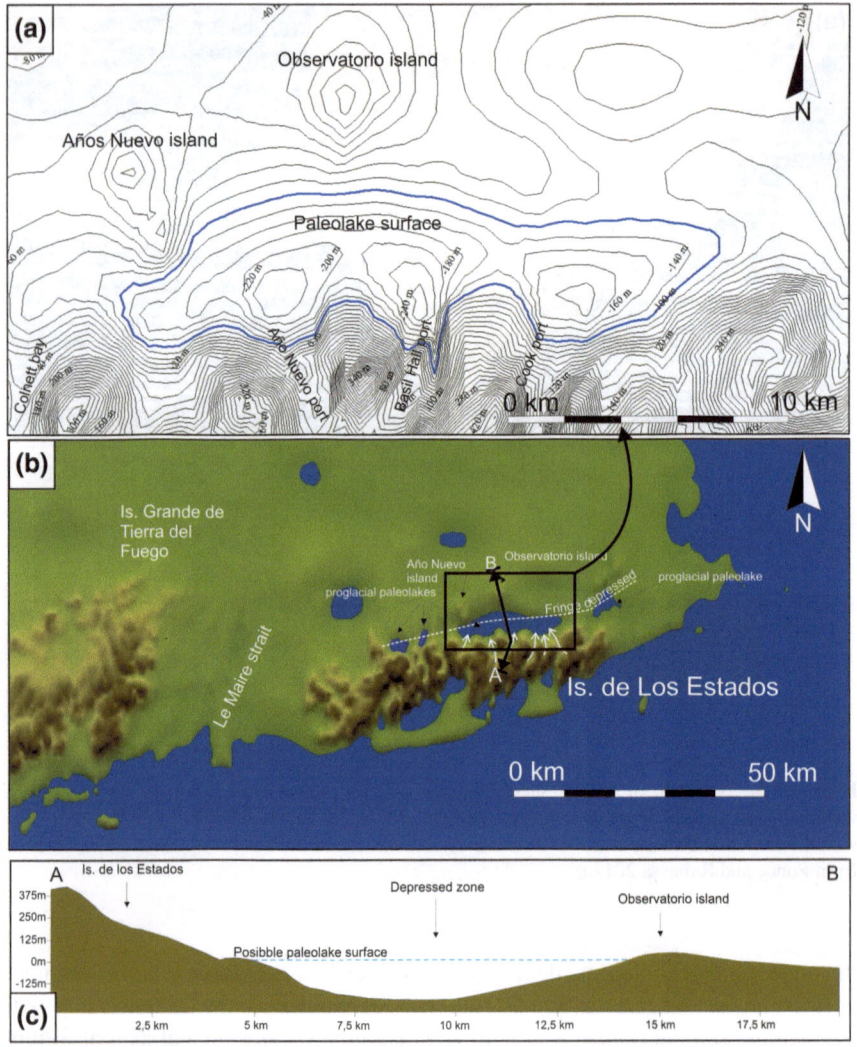

Fig. 5.18 a Depressions the product of the erosive action of the glaciers developed in the northern sector of Isla de los Estados. **b** Position of the sea level during the LGM. **c** Topographic profile showing a depression of over-excavation and possible paleolake

The absence of cirques associated with large troughs, as is the case of Puerto Cook and the troughs A6 and A6′, indicate the possibility of the accumulation of snow in the forms of ice sheets, with existing zones of accumulation independently from the cirques. According to what is observed, these troughs lack associated cirques and in some cases the cirques appear disproportioned in relation to the enormous troughs associated with them. The heights of the bases of these surfaces of accumulation are close to the current sea level (for example, Puerto Cook, 70 m a.s. l.). This form of accumulation of snow would demonstrate, consequently, that Isla de los Estados would have been completely covered by an ice sheet during these glacial events. From the analysis of the bases of the cirques on the island it can be estimated that the height of the equilibrium line or permanent snow line during the glaciations that affected the island was situated below 270 m a.s.l.(average height of the bases of the cirques). The value of R^2 of the line of regression calculated for the height of the bases of the cirques is not significant. One explanation for the high degree of dispersion in the heights of the cirques could be the fact that these landforms have not been generated during a single glaciation, but have been the result of various glacial cycles of different magnitudes. Rabassa et al. (1990a) estimated for the Last Glacial Maximum in the area of Punta Moat (Beagle Channel) a maximum height for the line of equilibrium close to 100–150 m a.s.l. judging from the position of lateral moraines developed at that locality. The absence of true lateral moraines on Isla de los Estados would indicate that during the LGM the height of the permanent snow line would have been located very close to or below current sea level.

The glaciers developed on Isla de los Estados would have had a thickness close to 300 m in the central axis of their flow (Ponce et al. 2009). This value was esti-mated adding the mean value of the maximum depths of the fjords and the mean height of the break in slope of the walls of the principal troughs and fjords.

The action of cryoturbation during the cold episodes after the glaciations on Isla de los Estados would have masked a large part of the original glacial land-forms. This would indicate the presence of a periglacial zone on the island that would have been present during most of the cold episodes that followed the glacial events. These processes are observed to be active today in the elevated zones of the eastern portion of the island, as is shown by the periglacial features described in San Juan del Salvamento, developed at higher than 500 m a.s.l. and composed by stone stripes and sorted circles (Ljung and Ponce 2006).

References

Ahlmann HW (1919) Geomorphological studies in Norway. Geogr Ann 1(1–148):193–252
Andrews J (1975) Glacial systems. An approach to glaciers and their environments. Duxbury Press, North Scituate, p 208
Auer V (1970) The pleistocene of fuego-patagonia. Part V: quaternary problems of Southern South America. Ann Acad Scientiarum Fennicae A III Geol-Geogr 100:1–194
Bermúdez FL, Recio JMR, Cuadrat JM (1992) Geografía física. Ediciones Cátedra, Madrid, p 594
Boelhowers J, Holdess S, Summer P (2003) The maritime subantarctic: a distinct periglacial environment. Geomorphology 52:39–55
Caminos R, Nullo F (1979) Descripción Geológica de la Hoja 67 e, Isla de los Estados. Territorio Nacional de Tierra del Fuego, Antártida e Islas del Atlántico Sur. Servicio Geológico Nacional. Boletín 175, Buenos Aires, p 52
Charlesworth JK (1957) The quaternary era. Edward Arnold, London, p 1700
Corte A (1977) Geocriología. El frío en la Tierra. Ediciones Culturales de Mendoza, Mendoza
Embleton C, King CA (1968) Glacial and periglacial geomorphology. Edward Arnold Ltd., London, p 608
Fairbridge R (1968) The encyclopedia of geomorphology. Encyclopedia of earth sciences series, vol 3. Reinhold Books Corp, Nueva York, p 1295

Flint R (1957) Glacial and pleistocene geology. Wiley and Sons, New York, p 553

Flint R, Fidalgo F (1964) Glacial geology of the east flank of the argentine andes between latitude 39°10′ S and latitude 41°20′ S. Geol Soc Am Bull 75:335–352

García MC (1986) Estudio de algunos rasgos geomorfológicos de la Isla de los Estados. Unpublished graduation thesis, Universidad Nacional del Centro de la Provincia de Buenos Aires and Centro Austral de Investigaciones Científicas (CADIC), Ushuaia, p 53

Garleff K (1977) Höhenstufen der argentinischen anden in cuyo, patagonien und feuerland. Göttinger Geogr Abh 68:1–150

Goldthwait RP (1976) Frost sorted material: a review. Quatern Res 6:27–35

Haynes V (1995) Alpine valley heads on the antarctic peninsula. Boreas 24:81–94

Heusser CJ (1989a) Late quaternary vegetation and climate of tierra del fuego. Quatern Res 31:396–406

Heusser CJ (1989b) Polar perspective of late quaternary climates in the southern hemisphere. Quatern Res 32:60–71

Klemsdal T (1982) Coastal classification and the coast of norway. Nor Geogr Tidsskr 36:129–152

Ljung K, Ponce JF (2006) Periglacial features on Isla de los Estados, Tierra del Fuego, Argentina. III Congreso Argentino de Cuaternario y Geomorfología. Actas, Córdoba, 1, 85–90

Markgraf V (1993) Palaeoenvironments and paleoclimates in Tierra del Fuego and southernmost Patagonia, South America. Palaeogeogr Palaeoclimatol Palaeoecol 102:53–68

Martínez O (2002) Geomorfología y geología de los depósitos glaciarios y periglaciarios de la región comprendida entre los 43° y 44° lat. Sur y 70° 30′ y 72° long. Oeste, Chubut, República Argentina. Unpublished Doctoral Thesis, Universidad Nacional de la Patagonia S.J.B., Esquel, p 327

Martini P, Brookfield ME, Sadura S (2001) Principles of glacial geomorphology and geology. Prentice Hall, Upper Sanddle River, p 381

Meierding TC (1982) Late pleistocene glacial equilibrium-line altitudes in the colorado front range – a comparison of methods. Quatern Res 18:289–310

Moller P, Hjort C, Björck S, Rabassa J, Ponce JF (2010) Glaciation history of Isla de los Estados, southeasternmost south America. Q Res 73(3) 521–534

Ponce JF, Rabassa J, Martínez O (2009) Morfometría y génesis de los fiordos de isla de los estados, tierra del fuego, argentina. Revista de la Asociación Geológica Argentina 65(4):638–647

Ponce JF, Rabassa J, Coronato A, Borromei AM (2011) Palaeogeographical evolution of the atlantic coast of pampa and patagonia from the last glacial maximum to the middle holocene. Biol J Linn Soc 103:363–379

Ponce JF, Rabassa J (2012a) Geomorfología glacial de la isla de los estados, tierra del fuego, argentina. Revista de la Sociedad Geológica de España 25(1–2):67–84

Ponce JF, Rabassa J (2012b) Historia de la plataforma submarina y la costa atlántica argentina durante los últimos 22.000 años. revista. Ciencia Hoy 22(127):50–56

Porter SC (1975) Equilibrium-line altitudes of late quaternary glaciers in the southern alps, New Zealand. Q Res 5:27–47

Rabassa J (2008) Late Cenozoic glaciations of Patagonia and Tierra del Fuego. In: Rabassa J (ed) Late Cenozoic of Patagonia and Tierra del Fuego. Develop in quat sci, Elsevier, Amsterdam, 11:151–204

Rabassa J, Serrat D, Martí C, Coronato A (1990) El Tardiglacial en el Canal Beagle, Tierra del Fuego, Argentina y Chile. XI Congreso Geológico Argentino, Actas 1, 290–293. San Juan, Argentina

Rabassa J, Coronato A, Roig C, Martínez O, Serrat D (2003) Un bosque sumergido en Bahía Sloggett, Tierra del Fuego, Argentina: Evidencia de actividad neotectónica en el Holoceno Tardío. Procesos geomorfológicos y evolución costera. Actas de la II Reunión de Geomorfología Litoral. Santiago de Compostela 2003: 333–345. Universidad de Santiago de Compostela publicacións

Rabassa J, Coronato A, Salemme M (2005) Chronology of the late cenozoic patagonian glaciations and their correlation with biostratigraphic units of the pampean region (Argentina). J South Am Earth Sci 20:81–103

Syvitski JPM, Shaw J (1995) Sedimentology and geomorphology of fjords. In Perillo GME (ed) Geomorphology and sedimentology of estuaries. Developments in sedimentology, vol, 53. Elsevier Science Publications, Amsterdam, p 113–178

Syvitski JPM, Burrel DC, Skei JM (1987) Fjords: processes and products. Springer, New York, p 379

Uriarte Cantolla A (2003) Historia del clima de la tierra. Victoria-Gasteiz: Servicio Central de Publicaciones del Gobierno Vasco, Bilbao, p 306

Valcárcel-Díaz M, Carrera-Gómez P, Coronato A, Castillo-Rodríguez F, Rabassa J, Pérez-Alberti A (2006) Cryogenic landforms in the sierras de alvear, fuegian andes, subantarctic argentina. Permafrost Periglac Process 17:371–376

Chapter 6
Paleogeography

Abstract In this chapter, we reconstruct the paleogeographic evolution of the Isla de los Estados for the period spanning the Last Glacial Maximum (LGM, ca. 24 cal. ka B.P.) to the present using the global sea-level rise curve proposed by Fleming et al. (1998) for this period and the Global Mapper 10 program. During the LMG, the Isla de los Estados was connected to the rest of the continent as the Isla Grande de Tierra del Fuego similarly did so. The northern coastline of the present Isla de los Estados, during this time, was found at around 100 km away from a straight line of its current position. The paleogeographic model has allowed to estimate the time of the opening of the Le Maire Strait, with the subsequent separation of Isla de los Estados and Isla Grande de Tierra del Fuego. This event would have taken place approximately at around 15 cal. ka B.P., when sea level rose above −85 m. Around 11,000 cal. years B.P., the small group of the Año Nuevo islands began to separate from the rest of Isla de los Estados, thus forming the archipelago with a configuration similar to the current arrangement.

Keywords Paleogeography • Isla de los Estados • Tierra del Fuego • Le Maire Strait • Last glacial maximum • Late glacial

6.1 Introduction

Isla de los Estados is located at the southern tip of the Argentine Continental Shelf (ACS). To the north, the island limits directly with the ACS, and toward the south, it is bounded by the continental slope. The Argentine Continental Shelf is located immediately east of the Argentine Atlantic coast. It extends approximately between latitudes 34° 00′ S/54° 20′ S and longitudes 56° 25′ W/66° 05′ W. Toward the east, it is limited by the continental slope. The ACS has an approximate extension of 1,000,000 km², a maximum length of 2,300 km in a NNE-SSW direction,

J. F. Ponce and M. Fernández, *Climatic and Environmental History of Isla de los Estados,*   69
Argentina, SpringerBriefs in Earth System Sciences, DOI: 10.1007/978-94-007-4363-2_6,
© The Author(s) 2014

and a mean width of 440 km in an E-W sense, with a minimum close to 180 km nearby Isla de los Estados, Tierra del Fuego, and a maximum of 880 km, N of the Malvinas/Falklands islands. It shows a maximum depth of around −250 m immediately W of these islands, but most of it is less than 200 m deep. The ACS is characterized by a quite gentle slope (less than 0.5°) and very low internal relief. The ACS is one of the more extensive submarine platforms in the world, most of which has been repeatedly uncovered during glacial times and flooded again during glacial terminations. Although the slope is very smooth, it shows clear variations along different portions of the platform, between 1:100 and 1:10,000, approximately.

6.2 Methodology

For the reconstruction of the paleogeographic evolution of the Península Mitre and Isla de los Estados, for the period spanning since the Last Glacial Maximum (LGM, ca. 24 cal. ka B.P.; Rabassa 2008) to the present, we used the program Global Mapper 10 for the elaboration of a digital model of sea-level rise, taking into consideration the curve of global sea-level rise since the Last Glacial Maximum proposed by Fleming et al. (1998). The digital land elevation models of the Shuttle Radar Topography Mission (SRTM), W100S10. BATHYMETRY.SRTM, and W060S10.BATHYMETRY.SRTM., with a 1 × 1 km resolution pixel were analyzed with this program. With these tools, several, successive paleogeographic maps were drawn, showing the position of sea level, starting in the lowermost level during the LGM (−120 m) to present sea level. The approximate timing for each of these sea-level positions was obtained by means of the Fleming et al. (1998) curve. The indicated calibrated ages are following their data.

6.3 The Paleogeographical Model

During the LGM, sea level was probably stable around 120 and 140 m below present sea level (Fleming et al. 1998; Uriarte Cantolla 2003). Under these conditions, a very large portion of the ACS was exposed, generating an enormous coastal plain along Pampa and Patagonia (Fig. 6.1). This huge plain extended continuously northwards to the central portion of the Brazilian Atlantic coast (14° 20' S), including most of the presently known as South American Continental Shelf. It has been estimated that the total surface area of this flatland was approximately 1,365,000 km^2 (Ponce et al. 2011a), an area as large and Spain and France put together. The portion of this plain extending along the coasts of Pampa and Patagonia, from the present mouth of the Río de la Plata estuary to its southernmost end at Isla de los Estados, was of approximately 590,000 km^2, with a varying width between 490 km (at the latitude of the present city of Bahía Blanca, Buenos Aires Province) and 100 km (at Península Mitre, Isla Grande de Tierra del Fuego)

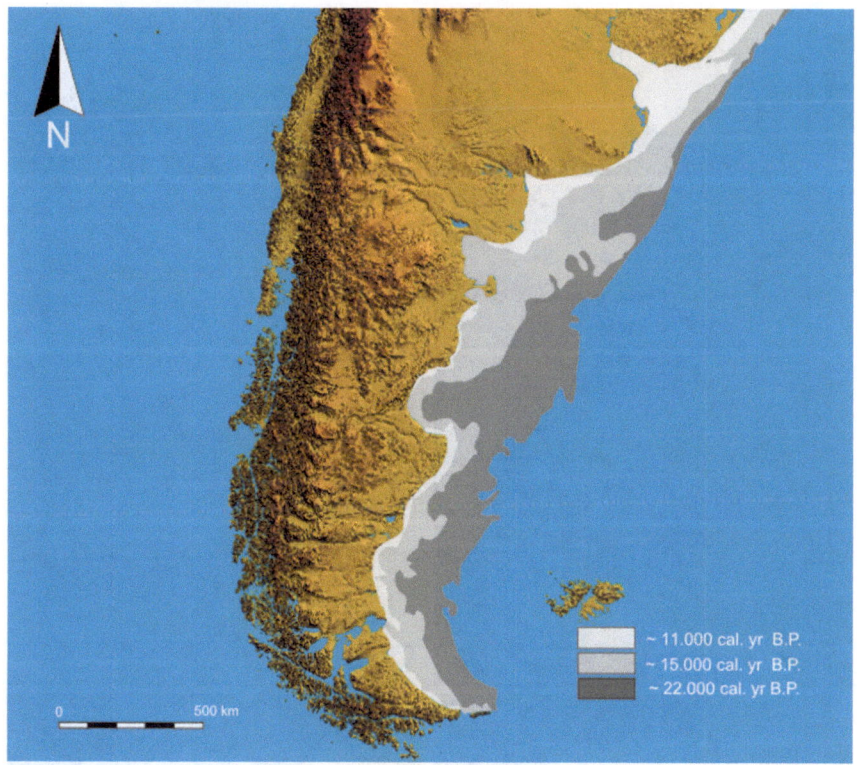

Fig. 6.1 Different positions of the coastline of Pampa and Patagonia during the LGM and the Late Glacial times. (Modified from Ponce et al. 2011a)

(Ponce et al. 2011a, b). The present Isla de los Estados was connected to the rest of the continent as the Isla Grande de Tierra del Fuego similarly did so. The northern coastline of the present Isla de los Estados, during the LGM, was found at around 100 km away from a straight line of its current position (Ponce et al. 2011a; Ponce and Rabassa 2012a). In contrast, the old line of the southern coast of the island was only 5 km to the south of its current position due to the presence of a very steep continental slope immediately to the south of the island (Ponce et al. 2009; Ponce and Rabassa 2012a) (Fig. 6.2a). The southern paleocoastal line of this island was, instead, located only at ca. 5 km S of its present position, due to the existence of a deep continental talus very close to the island. Under these conditions, the eastern boundary of the Patagonian continental area was extending ca. 450 km farther away from its present location. To the west, Isla de los Estados was directly joined to Isla Grande de Tierra del Fuego, whereas it was in contact with the open sea to the east.

Up to 17,000 cal. years B.P., Isla de los Estados remained united to the rest of the continent by means of a land bridge approximately 30 km in width (Fig. 6.2b).

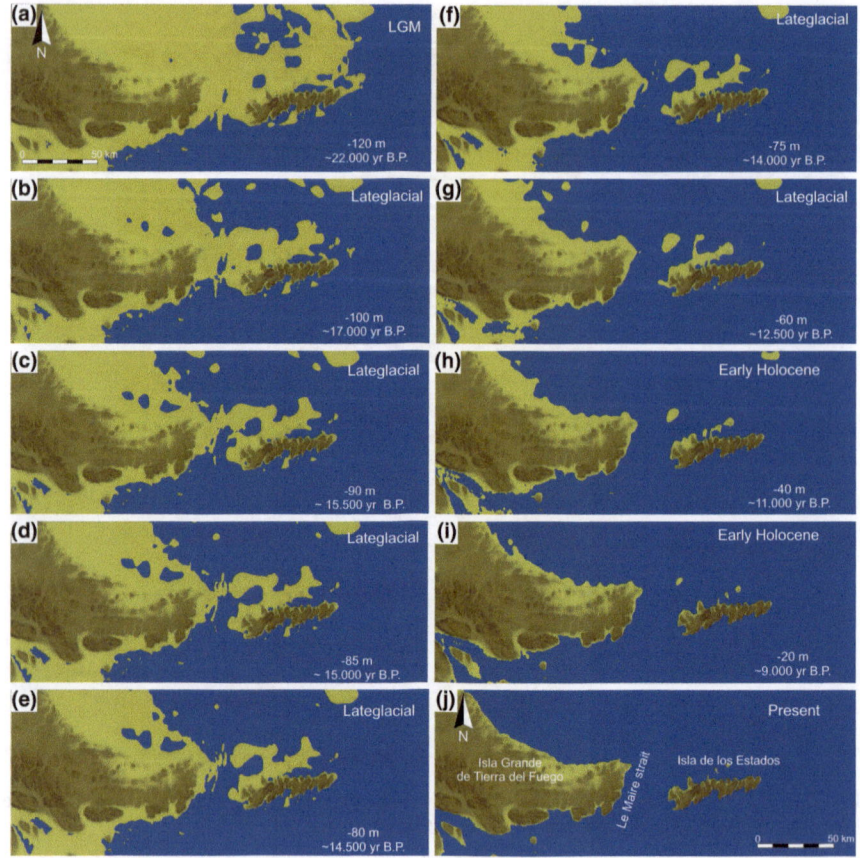

Fig. 6.2 Paleogeographic evolution of Península Mitre (SE of Isla Grande de Tierra del Fuego), Isla de los Estados, and the opening of the Le Maire strait (Modified from Ponce et al. 2011b)

Around this time, as a consequence of the sea-level rise, the large proglacial lakes that had developed on the northern coast of Isla de los Estados began to connect with the sea, forming a large bay, 25 km in length, on the NE coast of Isla de los Estados from San Juan del Salvamento to the eastern border of Colnett Bay. This bay remained as it was until around approximately 13,000 B.P.

Up to 15,500 cal. years B.P., the sea level had risen to 90 m below its current level and the extension of the great plain to the east of the present Argentine Atlantic coast had shrunk to approximately 65 % of its original area, now being only 400,000 km^2 in area (Ponce and Rabassa 2012b). On Isla de los Estados, the land bridge that connected the island with the rest of the continent had shrunk to a width of merely 8 km. The marine transgression across this bridge advanced from south to north, with the formation of a N–S oriented bay. Around this time, an exposed plain to the north of Isla de los Estados of approximately 1,200 km^2, equivalent to the double of its current surface area (Fig. 6.2c) still existed.

Some 500 years later, i.e., around 15,000 cal. years B.P., the definitive separation occurred between Isla Grande de Tierra del Fuego and Isla de los Estados, with the consequent formation of the Le Maire strait (Ponce et al. 2009), which had a minimum width of 700 m along a 6 km long canal (Fig. 6.2d). At the time of the separation of Isla de los Estados from the rest of the continent, the area of the island was approximately 1,600 km², i.e., to say three times its current area. Toward 12,500 cal. years B.P. the Le Maire strait reached its present width and the area of the island was equivalent to double the current surface area (900 km²).

Around 11,000 cal. years B.P., when sea level reached −40 m, the exposed plain had almost completely disappeared east of Patagonia, with remnants of significant extension principally remaining to the east of the current coastline of the province of Buenos Aires (175 km width in the Bahía Blanca area and 380 km to the east of the Río de la Plata estuary). During this time, in southern Patagonia, principally along the current coast of Santa Cruz province, the extension of the paleoplain did not exceed 25 km in width (Ponce et al. 2011a). Isla de los Estados had then an area of approximately 730 km² (Fig. 6.2h). Around this time, the small group of the Año Nuevo islands began to separate from the rest of Isla de los Estados, thus forming the archipelago with a configuration similar to the current arrangement.

The separation of Isla Grande de Tierra del Fuego from the rest of the continent and the formation of the present Magellan Strait occurred around 10,200 cal. years B.P. when sea level rose to −35 m (Ponce 2009; Ponce et al. 2011a).

The last remnants of the exposed flatland survived until 9,000 cal. years B.P., which is when the Argentine coasts acquired their current configuration, as did Isla de los Estados (Fig. 6.2i).

References

Fleming K, Johnston P, Zwartz D, Yokoyama Y, Lambeck K, Chappell J (1998) Refining the eustatic sea-level curve since the Last Glacial Maximum using far- and intermediate-field sites. Earth Planet Sci Lett 163:327–342

Ponce JF (2009) Palinología y geomorfología del Cenozoico tardío de la Isla de los Estados. Unpublished Doctoral Thesis, Universidad Nacional del Sur, Bahía Blanca, p 192

Ponce JF, Rabassa J, Martínez O (2009) Fiordos en Isla de los Estados: descripción morfométrica y génesis de los únicos fiordos en la Patagonia Argentina. Revista de la Asociación Geológica Argentina 65(4):638–647

Ponce JF, Rabassa J, Coronato A, Borromei AM (2011a) Paleogeographic evolution of the Atlantic coast of Pampa and Patagonia since the Last Glacial Maximum to the Middle Holocene. Biol J Linn Soc 103:363–379

Ponce JF, Borromei AM, Rabassa J (2011b) Evolución del paisaje y la vegetación durante el Cenozoico tardío en el extremo sureste del Archipiélago Fueguino y Canal Beagle. In: Zangrando AF, Vázquez M, Tessone A (eds) Los cazadores-recolectores del extremo oriental fueguino. Arqueología de Península Mitre e Isla de los Estados. Sociedad Argentina de Antropología, Buenos Aires, pp 31–64

Ponce JF, Rabassa J (2012a) Geomorfología glacial de la Isla de los Estados, Tierra del Fuego, Argentina. Revista de la Sociedad Geológica de España 25(1–2):67–84

Ponce JF, Rabassa J (2012b) Historia de la Plataforma Submarina y la Costa Atlántica Argentina durante los últimos 22.000 años. Revista Ciencia Hoy 22, 127:50–56. Buenos Aires

Rabassa J (2008) Late Cenozoic glaciations of Patagonia and Tierra del Fuego. In: Rabassa J (ed) Late Cenozoic of Patagonia and Tierra del Fuego, vol 11. Developments in Quaternary Science, Elsevier, pp 151–204

Uriarte Cantolla A (2003) Historia del clima de la tierra. Servicio Central de Publicaciones del Gobierno Vasco, Victoria-Gasteiz, p 306

Chapter 7
Palinology

Abstract A vegetation reconstruction was made in western Isla de los Estados using palynological analysis for the last 13,000 cal year BP. The pollen data indicates initial treeless herbaceous and paludal vegetation with scarce *Empetrum*/Ericaceae type heaths and scrubs as a result of plan invasion and short-term succession vegetal communities spreading over the shoreline areas under locally more humid conditions. The vegetation between 10,300 and 8300 cal year BP included dwarf shrub heaths, scrubs, cushion plants and grasses with scattered trees, under warmer and drier climate conditions than today. After 8300 cal year BP, more humid conditions allowed the expansion of an open *Nothofagus* forest associated with dwarf shrub heath communities. It was followed at 6700 cal year BP by a gradual closed forest development in association with *Drimys winteri* and shrub and herb vegetation indicative of Subantarctic Evergreen Forest-Magellanic Moorland vegetational transition. After 5500 cal year BP, the rate of evergreen beech forest greatly increased with the development of almost pure Subantarctic Evergreen Forest communities. These vegetational changes accompanied a modification of the climate toward colder and wetter conditions. After 2700 cal year BP, the closed forest was replaced by an open *Nothofagus* forest indicative of warmer and drier conditions.

Keywords Palinology • Isla de los Estados • Tierra del Fuego • Subantarctic evergreen forest • Pollen • Vegetation • Late glacial • Holocene

7.1 Introduction

Isla de los Estados, due to its geographical location, conforms a unique and sensitive area for Quaternary paleoecological and paleoclimatic studies, providing information on the atmospheric and environmental conditions from cold-temperate high latitudes in the Southern Hemisphere.

J. F. Ponce and M. Fernández, *Climatic and Environmental History of Isla de los Estados,*   75
Argentina, SpringerBriefs in Earth System Sciences, DOI: 10.1007/978-94-007-4363-2_7,
© The Author(s) 2014

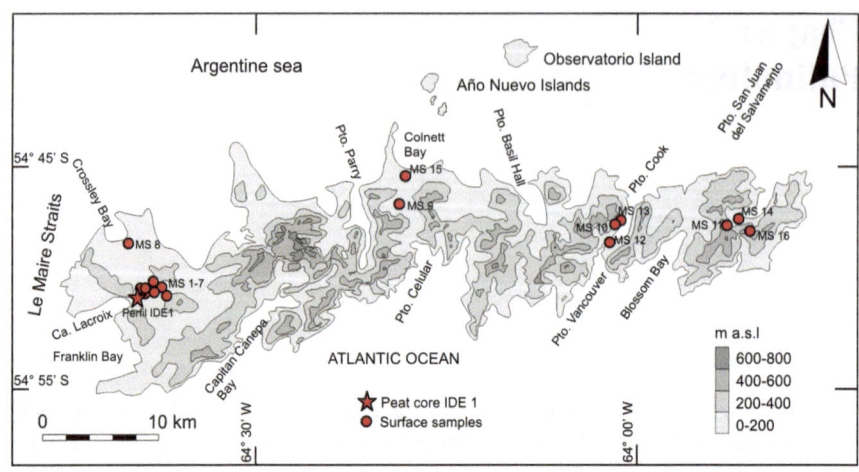

Fig. 7.1 Map of Isla de los Estados showing the location of IDE-1 peat core and surface samples

Only three studies report on the Quaternary palinology of Isla de los Estados. The first one is a palynological study by Johns (1981), based on three peat cores collected in 1971. No absolute dates are available for these sequences. The second study (Björck et al. 2012) analyzed an almost 14,000 year old peat sequence located on the northern part of the island based on a variety of methods that included pollen and spore analysis. This chapter transcripts the most important aspect of the third of these works (Ponce et al. 2011) which was based on the pollen analysis of a peat bog (IDE-1 pollen section) located at Caleta Lacroix (Bahía Franklin, western Isla de los Estados, Figs. 7.1 and 7.2) and the geomorphological analysis of the area to infer the paleoenvironmental and paleoclimatic conditions during the last 13,000 years.

7.2 Materials and Methods

The IDE-1 pollen section (54° 50' 46,5" S; 64° 38' 50,8" W; 27 m a.s.l.) is located in an interdune area. Between the base at 330 and 155 cm depth, the core consists of highly humified dark brown peat with fibrous. Between 155 and 110 cm depth, compact highly humified peat and dark brown sandy sediments are found. Dark compact peat layers interspersed with dark grey sand lenses are present between 110 and 75 cm depth. The upper 75 cm consists of dark brown to bluish gray brown non-humified fibrous peat.

The fossil-peat core IDE-1 was taken with a Russian corer (Fig. 7.3). In the laboratory, the core was sub-sampled at 5 cm intervals and the sediments were described. A total of 65 fossil pollen samples were obtained. In order to obtain modern pollen data for the interpretation of the paleovegetation changes from the fossil pollen spectra, 16 surface samples were extracted from the studied area and northern coast of the island (Fig. 7.1). Modern pollen frequencies are plotted in Fig. 7.4.

Fig. 7.2 Caleta Lacroix, Bahía Franklin, western Isla de los Estados (Photo: Ponce 2010)

Fig. 7.3 Sampling with a Russian corer (Photo: Ponce 2003)

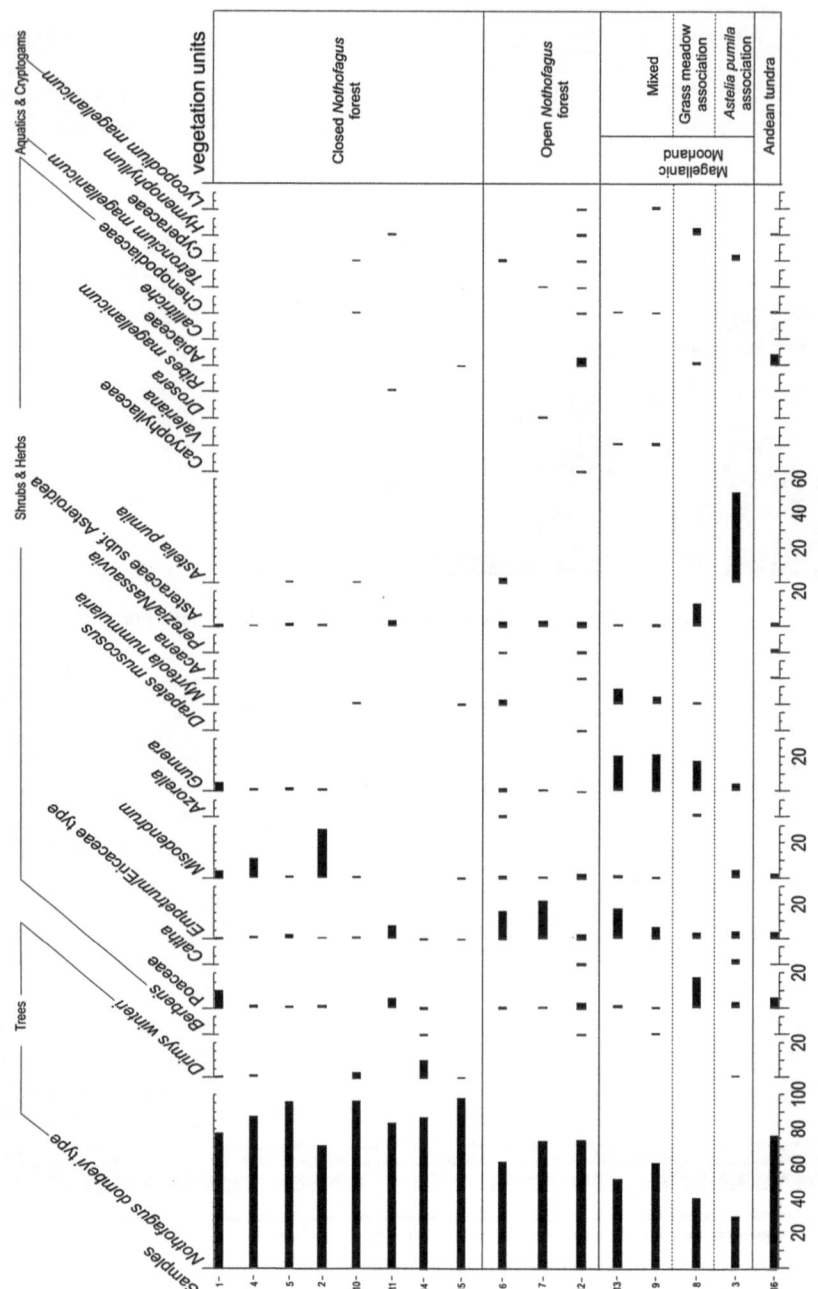

Fig. 7.4 Surface pollen frequency diagram (%). Surface samples collected at sites are indicated by number (Ponce et al. 2011a)

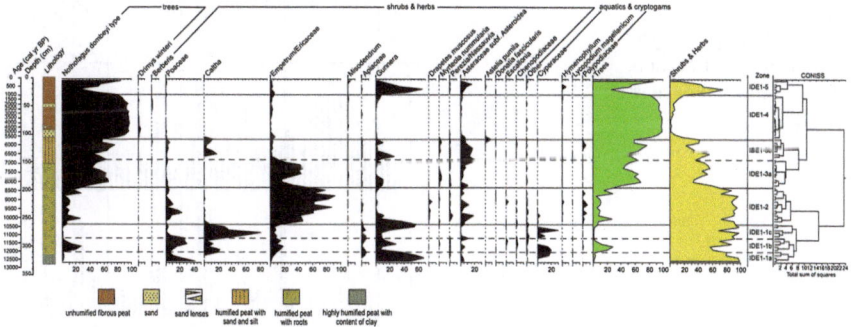

Fig. 7.5 Fossil pollen/spore frequency diagram (%) and stratigraphy at IDE-1 section, Caleta Lacroix. (Modified from Ponce et al. 2011a)

For the pollen analysis, peat samples were prepared according to standard Faegri and Iversen (1989) techniques. *Lycopodium* spore tablets added to each sample prior to treatment (Stockmarr 1971) allowed us to calculate the pollen concentration per gram of sediment. Frequencies (%) of tree, shrub and herb pollen of terrestrial origin were calculated from sums mostly of ≥ 300 grains. Pollen of aquatic plants and cryptogams were calculated separately and related to the sum of terrestrial pollen. Fossil frequency and concentration pollen data are plotted in Figs. 7.5 and 7.6. Other herbs include taxa with low values, such as Caryophyllaceae, Rubiaceae, *Valeriana*, Scrophulariaceae, *Azorella*, *Acaena*, *Pratia*, *Cardamine*, *Rubus*, *Crassula*, Saxifragaceae, Onagraceae, *Geum* and Solanaceae. Using the Cavalli-Sforza Distance (TGView 2.0.2, Grimm 2004), a stratigraphically constrained cluster analysis was applied to distinguish pollen zones considering taxa that reach percentages of ≥ 1 % of the sum of terrestrial pollen.

Pollen from the evergreen species *N. betuloides* and the deciduous species *N. pumilio* and *N. antarctica* are reported as the "*Nothofagus dombeyi* type", given the difficulty in the identification of the different species. Another special case is *Empetrum rubrum*, *Gaultheria/Pernettya* (Ericaceae) and *Lebetanthus myrsinites* (Epacridaceae) which are morphologically similar and occur as tetrads; for these reasons, they are considered as one taxonomic group on the pollen diagrams in the present study, named "*Empetrum*/Ericaceae type".

Five peat samples provided chronologic control for the peat section. The NSF-Arizona AMS Laboratory (U.S.A.), undertook the radiocarbon analysis on the samples, and the radiocarbon ages were converted to calendar years BP using the program CALIB 6.0 (Stuiver et al. 2005) and the Southern Hemisphere curve (SHCal04) (McCormac et al. 2004) (Table 7.1).

Fig. 7.6 Fossil pollen concentration (grains/gr) diagram at IDE-1 section, Caleta Lacroix. (Ponce et al. 2011a)

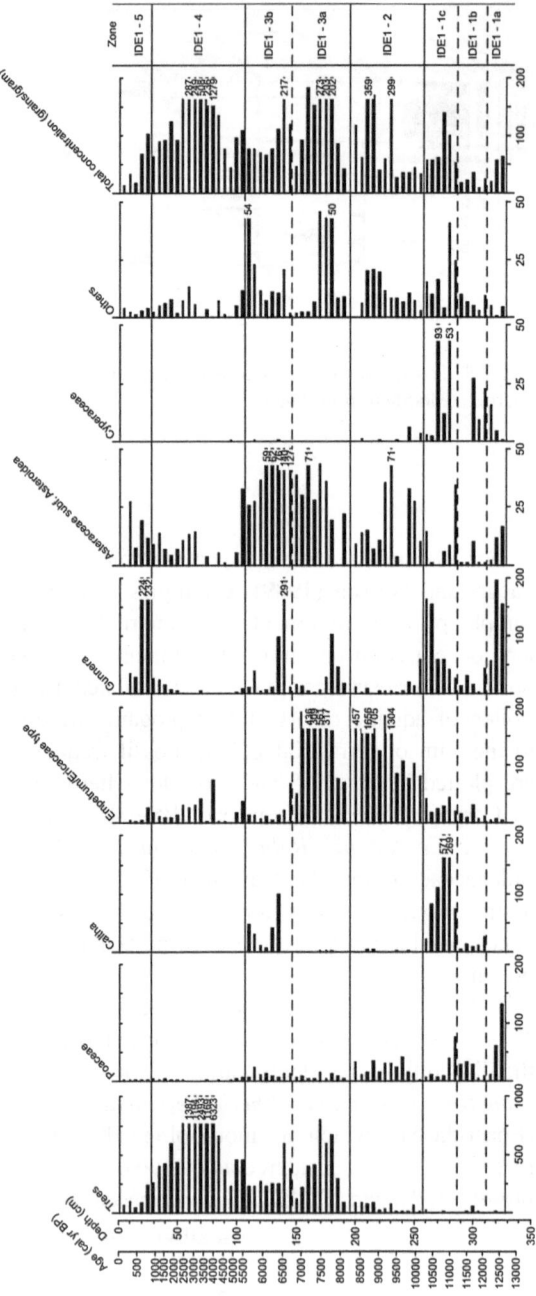

Table 7.1 AMS ^{14}C dates and calibrated ages of selected samples from IDE-1 section, Caleta Lacroix, western Isla de los Estados

Depth (m)	Laboratory No	^{14}C year BP	Calibrated years BP (median probability)	1σ range	2σ range	Material
0.30	AA72174	1081 ± 34	990	955–1004	933–1056	Peat
1.05	AA75398	4811 ± 49	5525	5476–5540	5464–5645	Peat
2.00	AA72175	7645 ± 45	8434	8390–8457	8382–8540	Peat
2.60	AA75399	9174 ± 51	10,340	10,313–10,394	10,235–10,439	Peat
3.30	AA62509	10,670 ± 39	12,607	12,566–12,636	12,548–12,686	Clay gyttja

7.3 Results

7.3.1 Modern Pollen Data

The pollen spectra from surface samples (Fig. 7.4) represent the principal units of vegetation found in the island: *Nothofagus* forest, Magellanic Moorland and Andean vegetation. The pollen concentration values of surface samples are not shown in Fig. 4.1.

1. Closed *Nothofagus* forest: characterized by the highest percentages of *Nothofagus dombeyi* type (73–98 %). The data reveal the relative importance of *Drimys winteri* (9 %) and *Misodendrum* (27 %) and the lowest values of *Empetrum*/Ericaceae type. Poaceae with values of 10 % are related to grass meadows close to the sampled site. Total pollen concentration values are high (1,746,000–51,000 grains/gr) contributed mainly by *Nothofagus dombeyi* type (up to 825,000 grains/gr). Relevant quantities of *Drimys winteri* pollen indicate the presence of Sub-Antarctic Evergreen Forest (*Nothofagus betuloides-Drimys winteri* association) that extends toward the outer coastal zone of heavy precipitation (800-> 1000 mm/year; Pisano 1977). Records of the beech parasite *Misodendrum* (27 %) suggest local establishment of forest elements.
2. Open *Nothofagus* forest: dominance of *Nothofagus dombeyi* type pollen characterized this vegetation unit with values up to 74 %, accompanied by *Empetrum*/Ericaceae type (22 %) and great diversity of herbs and shrubs. In this forest, such species as *Empetrum* may become locally sub-dominant in the more open facies (Moore 1983). Total pollen concentration values are lower than of the previous unit varying between 600,000 and 46,300 grains/gr. The *Nothofagus* concentration records a maximum of 226,600 grains/gr.
3. Magellanic Moorland: this vegetation unit is represented by three different bog communities: mixed, grass meadow and *Astelia pumila* associations. In general *Nothofagus dombeyi* type reaches pollen values up to 60 %. Peaks of *Empetrum*/Ericaceae type (21 %), *Gunnera* (21 %) and *Myrteola nummularia* (8 %) reflect the character of mixed bog communities. High values of *Astelia pumila* (52 %) characterize the most notable community of this unit,

which probably is fed by an annual precipitation of more than about 2000 mm and comprises a dense low covering cushion plants. Grass meadow association shows the relative importance of Poaceae (17 %), *Gunnera* (17 %) and Asteraceae subf. Asteroideae (12 %). The latter association represents a distinctive grassland vegetation that occurs on sheltered coastal localities, which show a markedly oceanic influence (Moore 1983). Total pollen concentration values recorded in this unit range between 127,000 and 10,200 grains/gr.

4. Andean tundra: the vegetation of this unit lies at and just above timberline (450 m a.s.l.). The Andean communities consist of sparse and scattered plants that occur mainly in sheltered hollows and peat pockets between rocks (Dudley and Crow 1983). Increased quantities of *Nothofagus dombeyi* type pollen (up to 75 %) are indication of great atmospheric dispersion at this altitudinal site. The presence of "krummholz" stands formed by *Nothofagus antarctica* specimens at the altitudinal limit of forest are also inferred. Cushion and heath plants such as *Azorella* (6 %) and *Empetrum*/Ericaceae type (3 %) are present. Close associates include Poaceae, *Acaena*, Asteraceae subf. Asteroideae and *Nassauvia*. The total pollen concentration reaches 352,000 grains/gr with the highest values corresponding to *Nothofagus dombeyi* type (134,000 grains/gr).

7.3.2 Fossil Pollen Data. IDE-1 Section

Cluster analysis recognized five main zones: IDE1-1 to IDE1-5 based on conspicuous changes in the pollen stratigraphy (Fig. 7.5). In order, from the lower to the upper part of the section, they are:

Zone IDE1-1 (330–260 cm, ca.12,600–10,300 cal yr BP)

The non-arboreal pollen dominates the zone with high values (57–98 %) whereas the arboreal pollen displays low values (<27.7 %). Three subzones can be differentiated based on changes in the proportions of non-arboreal and arboreal taxa. The initial subzone (IDE1-1a: 12,600–12,000 cal year BP) is dominated by *Gunnera* (36–68 %) and Poaceae (6–42 %). The middle subzone (IDE1-1b: 12,000–11,200 cal year BP) shows an increase in *Nothofagus dombeyi type* (28 %), *Caltha* (22 %) and *Empetrum*/Ericaceae type (18 %), percentages also accompanied by Poaceae (30 %) and *Gunnera* (29 %). The upper subzone (IDE1-1c: 11,200–10,300 cal yr BP) displays an increase in *Caltha* and *Gunnera* percentages (up to 82 and 57 % respectively), whereas *Nothofagus dombeyi* type decreased to 7 % and Poaceae to 8 %. Cyperaceae are present, although fluctuating, with peaks of 30 %. Total pollen concentration (Fig. 7.6) is moderate (<650,000 grains/gr) in the subzone IDE1-1a, contributed mainly by *Gunnera* (195,000 grains/gr) and Poaceae (132,000 grains/gr). A decline in total values occurs in subzone IDE1-1b (354,000 grains/gr), with an arboreal concentration peaks of 49,000 grains/gr accompanied by Poaceae, *Empetrum*/Ericaceae type and *Gunnera* (~32,000 grains/gr). Within subzone IDE1-1c, *Gunnera* dominates, achieving a maximum of 571,000 grains/gr and contributing to the bulk of the total concentration values (1,400,000 grains/gr).

Zone IDE1-2 (260–195 cm, 10,300–8300 cal yr BP)

Non-arboreal pollen maintains high percentages (95.8 %). The zone features the dominance of *Empetrum*/Ericaceae type (43–92 %), accompanied by *Nothofagus dombeyi* type (28 %), Poaceae (23 %) and Asteraceae subf. Asteroideae (19 %). Increasing total pollen concentration values, as high as 3,590,000 grains/gr, are caused primarily by *Empetrum*/Ericaceae type (1,650,000 grains/gr) and secondarily by *Nothofagus dombeyi* type (91,500 grains/gr) (Fig. 7.6).

Zone IDE1-3 (195–105 cm, 8300–5500 cal yr BP)

This zone shows the co-dominance of arboreal and non-arboreal taxa. Two subzones can be identified on the basis of their proportional changes. Subzone IDE1-3a (8300–6700 cal yr BP) is dominated by *Nothofagus dombeyi* type (41–66 %) and *Empetrum*/Ericaceae type (11–47 %) accompanied by low proportions of Asteraceae subf. Asteroideae (16 %) and *Gunnera* (11 %). Subzone IDE1-3b (6700–5500 cal yr BP) shows a decrease in *Empetrum*/Ericaceae type values to 3 % and increase in *Nothofagus dombeyi* type values (up to 76 %) accompanied by *Gunnera* (27 %), Asteraceae subf. Asteroideae (21 %) and *Caltha* (18 %). *Drimys winteri* appears in low values (0.8 %) but becomes continuous throughout the sequence. Total pollen concentration recorded higher values during subzone IDE1-3a (2,730,000 grains/gr) and decreased in subzone IDE1-3b (662,000 grains/gr) (Fig. 7.6).

Zone IDE1-4 (105–30 cm, 5500–1000 cal yr BP)

The arboreal taxa dominate this zone. *Nothofagus dombeyi* type exhibits the highest values (97 %) in the entire peat core accompanied by *Drimys winteri* that also records its maximum values (3 %). Total pollen concentration (Fig. 7.6) values range between 12,792,000 and 450,000 grains/gr. *Nothofagus dombeyi* type gains prominence between 4000 and 2700 cal yr BP, achieving a maximum of 6,323,000 grains/gr.

Zone IDE1-5 (30–0 cm, 100–00 cal yr BP)

Non-arboreal taxa dominate the zone. The record of *Nothofagus dombeyi* type suddenly drops to 25 % and later, climbs up to 86 % towards the end of the zone. Also, *Gunnera* frequency rises to 66 % and later, declines to 4 %. Close associates include Asteraceae subf. Asteroideae (17 %). Total pollen concentration values (Fig. 7.6) decrease abruptly reaching a minimum of 142,000 grains/gr. *Nothofagus* records 45,300 grains/gr and *Gunnera* peaks at 232,000 grains/gr.

7.4 Vegetation Reconstruction

Initial pollen assemblage (Fig. 7.5) shows the dominance of pioneer herbs like *Gunnera* and Poaceae, spreading into the treeless lowland areas (subzone IDE1-1) between 12,600 and 10,300 cal yr BP (Fig. 7.7a). The increase of hydric *Caltha* indicates a probable shift toward locally more humid environments during subzone IDE1-1c. Meanwhile, the record of Cyperaceae suggests minerotrophic conditions and water availability. The vegetation, in addition, would have included heath (*Empetrum*/Ericaceae type) and scrub (Asteraceae subf. Asteroideae) communities. Presence of shrubby littoral vegetation related to present shoreline areas in sheltered bays (Dudley and Crow 1983), also can be seen by minor but significant occurrence of *Escallonia* pollen (subzone IDE1-1b). All these communities are developed in locally wet settings such as flat-lying areas of slow-moving and impeded drainage, possible affected by snowmelt runoff from local glacier meltwater discharge, and reflect a variety of succession sequences on these deglacial landscapes following glacier recession. The low frequency and concentration values of *Nothofagus* (Fig. 7.6) imply sources from extra-local forest refuges and/ or low pollen production due to unsuitable environmental conditions for tree growth. Compared with minima in the early part of the record, *Nothofagus* frequencies in subzone IDE1-1b are higher (28 %) (Fig. 7.5). This assemblage (subzone IDE1-1b) dated between 12,000 and 11,200 cal yr BP is possibly the result of short-term plant invasion from some proximal forest stand or westerly wind intensification that favored the long-distant atmospheric transport from distant sites.

Between 10,300 and 8300 cal yr BP *Nothofagus* trees gradually started to spread in the landscape (Fig. 7.7b). It is probable that *N. antarctica* were involved in the processes of colonization first, occupying those sites less favorable to growth. According to modern analogues in southern Chile, *N. antarctica* is a pioneer species that colonizes the deglaciated land first (Donoso 1993, in Fesq-Martin et al. 2004). It has wider ecological amplitude than *N. pumilio* and is prominent

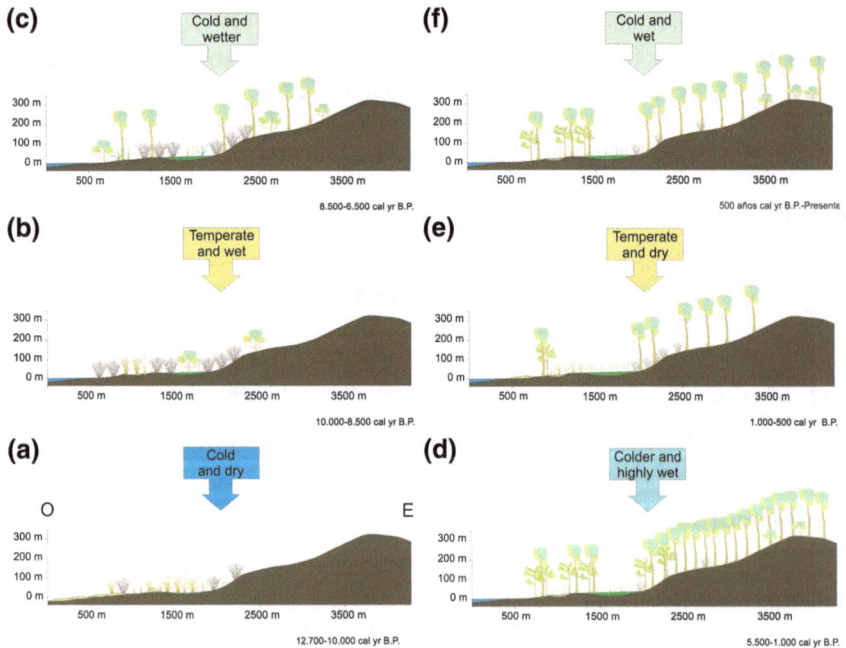

Fig. 7.7 Vegetation reconstruction at Bahia Fanklin during the Late Glacial-Holocene times

where shallow soils, high water table and aridity prevail (Moore 1983). Close associates of the more open forest communities are the occurrence of the parasite *Misodendrum* and the epiphitic *Hymenophyllum* fern that grow on the forest floor (Dudley and Crow 1983). During this interval, the expansion of dwarf shrub heath (*Empetrum*/Ericaceae type) communities (zone IDE1-2) indicates low exchange capacity and pH conditions on the mire (Fig. 7.5). *Empetrum* is a dominant species that develops over acid, shallow and well-drained soils (Moore 1983) in association with *Drapetes muscosus* and *Myrteola nummularia* (Moore 1983). Openness in landscape is also inferred by the record of light-dependent Polypodiaceae ferns (Heusser 1998). A comparable plant assemblage can be found today in Fuego-Patagonian steppe where mean annual precipitation values not surpass 400 mm/year and summer temperatures averaging between 11 and 12°C (Pisano 1977). This evidence suggests an increase in temperature and effective moisture, in spite of precipitation which continued lower than today.

An abrupt rise of arboreal taxa occurred between 8300 and 5500 cal yr BP, as indicated by high percentage and concentration of *Nothofagus* pollen values (zone IDE-3) (Figs. 7.5 and 7.6). The arboreal pollen show frequent, large-amplitude fluctuations that indicate high variability in forest cover near the studied site. Between 8300 and 6700 cal yr BP (subzone IDE-3a), the landscape displays a physiognomy of a closed *Nothofagus* forest interspersed with dwarf shrub heath (*Empetrum*/Ericaceae type, *Myrteola nummularia*), scrub (Asteraceae subf. Asteroideae) and herb (*Gunnera*, Poaceae) communities related to increased effective precipitation (Fig. 7.7c). Analogous communities exist today in the Sub-Antarctic Deciduous Forest in connection with *Sphagnum* bogs and their setting at lower altitudes from southern Isla Grande de Tierra del Fuego where mean annual precipitation totals between 500 and 800 mm and mean annual temperature averages 7°C (Pisano 1977; Heusser 1998).

After 6700 cal yr BP (subzone IDE-3b), the record of *Drimys winteri*, a typical associate of *Nothofagus betuloides*, along with an increase of *Caltha*, a drop of *Empetrum*/Ericaceae type and a record of cushion bog (*Astelia pumila*, *Myrteola nummularia*) and prostate dwarf shrub (*Berberis*, Asteraceae subf. Asteroideae) communities indicate the establishment of Sub-Antarctic Evergreen Forest-Magellanic Moorland transition (Fig. 7.7d) (Crow and Dudley 1983; Moore 1983). The development of this vegetation unit implies further increase in precipitation that culminated with the establishment and persistence of closed-canopy forest communities of Sub-Antarctic Evergreen Forest between 5500 and 1000 cal yr BP (Fig. 7.7e). The highest concentration values of *Nothofagus* pollen similar to those of the surface samples from the closed *Nothofagus* forest are recorded between 4000 and 2700 cal yr BP (Fig. 7.6) implying that precipitation reached its maximum level of the record. Modern manifestation of these communities are found in the extreme southeastern and outer coastal zone of the Fuegian Archipelago where mean annual precipitation is on the order of 800-> 1000 mm and mean annual temperature averages 6.5°C (Heusser 1998).

These highly humid and cool conditions ended when *Nothofagus* pollen declined abruptly after 2700 cal yr BP reaching minimum values at 500 cal yr BP (zone IDE1-5). The pollen record suggests a forest reduction and expansion of *Gunnera* and scrubs (Asteraceae subf. Asteroideae) in the area probably related to warmer and drier conditions (Fig. 7.7f). After 500 cal yr BP, the *Nothofagus* forest shows a recovery and the landscape displays a physiognomy of an open forest (Fig. 7.7g) as can be seen by the low tree concentration values (Fig. 7.6) similar to those of the surface samples from the open *Nothofagus* forest unit.

References

Birks HJB, Birks HH (1980) Quaternary Palaeoecology. Botany School, University of Cambridge, Edward Arnold pp 289

Björck S, Rundgren M, Ljung K, Unkel I, Wallin A (2012) Multy-proxy analysis of a peat bog on Isla de los Estados, easternmost Tierra del Fuego: a unique record of the variable Southern Hemisphere Westerlies since the last deglaciation. Quatern Sci Rev 42:1–14

Dudley TR, Crow GE (1983) A contribution to the Flora and Vegetation of Isla de los Estados (Staaten Island), Tierra del Fuego, Argentina. American Geophysical Union, Antarctic Research Series , Washington D. C, 37, 1–26

Faegri K, Iversen J (1989) Textbook of Pollen Analysis, 4th edn. Willey, Copenhague

Fesq-Martin M, Friedmann A, Peters M, Behrmann J, Kilian R (2004) Late-glacial and Holocene vegetation history of the Magellanic rain forest in southwestern Patagonia, Chile. Veg Hist Archaeobotanic 13:249–255

Grimm E (2004) Tilia and TGView 2.0.2. Illinois State Museum. Research and collection Center. Springfield, Illinois

Heusser CJ (1998) Deglacial paleoclimate of the American sector of the Southern Ocean: Late Glacial-Holocene records from the latitude of Canal Beagle (55° S), Argentine Tierra del Fuego. Palaeogeogr Palaeoclimatol Palaeoecol 141:277–301

Jhons WH (1981) The vegetation history and paleoclimatology for the Late Quaternary of Isla de los Estados. Argentina, USA

McCormac FG, Hogg AG, Blackwell PG, Buck CE, Higham TFG, Reimer PJ (2004) SHCal04 Southern Hemisphere Calibration 0–1000 cal BP. Radiocarbon 46:1087–1092

Moore MD (1983) Flora of Tierra del Fuego. Antony Nelson England, USA pp 369

Pisano E (1977) Fitogeografía de Fuego-Patagonia chilena. Comunidades vegetales entre las latitudes 52° y 56° S. Anales del Instituto de la Patagonia, Punta Arenas, 8, 121-250

Ponce JF, Borromei AM, Rabassa J, Martínez O (2011) Late Quaternary palaeoenvironmental change in western Staaten Island (54.5°S, 64°W) Fuegian Archipelago. Quatern Int 233:89–100

Stockmarr J (1971) Tablets with spore used in absolute pollen analysis. Pollen Spore 13:615–621

Stuiver M, Reimer P (1993) Extended 14C database and revised CALIB radiocarbon calibration program. Radiocarbon 35:215–230

Chapter 8
Diatom Analysis

Abstract In this chapter we present the diatom analysis results from glacio-lacustrine and lacustrine sedimentary sequences (Laguna Cascada, 54° 45′51″ S, 64° 20′20.7″W and Lago Galvarne Bog, 54° 44′16″S, 64° 19′37.9″W). The deeper part of Laguna Cascada core was radiocarbon dated in 13.285 ± 80 [14]C B.P. (15.949–15.545 cal. years B.P.). Diatom analysis has showed the existence of significant climate and paleoenvironmental fluctuations around the end of the Late Glacial and beginning of the early Holocene. The deglacial period was dominated by fragilarioids species. However, diatom assemblages had changed during the Holocene due to different climate trends. Diatom analyses from Lago Galvarne Bog were carried out only in the section corresponding to the middle Holocene marine transgression. The base of the core was dated at 13,515 [14]C B.P. (16,260 cal. years B.P.). Between 8,000 and 7,400 cal. years B.P., the dominance of marine diatoms together with species of brackish waters has suggested a strong marine influence within the basin. Between 7,400 and 3,700 cal. years B.P., a coastal environment would have been developed at Lago Galvarne, or perhaps even a fjord-type environment, as suggested by the high frequency of brackish diatoms together with fresh waters.

Keywords Diatoms • Laguna Cascada • Lago Galvarne • Lateglacial • Holocene marine transgression

8.1 Introduction

Diatoms (Bacillariophyceae) are unicellular, photosynthetic algae that are extremely sensitive to physical (e.g., turbidity, temperature, light) and chemical (e.g., pH, nutrients, salinity) changes in lakes and other water bodies. The well preserved diatom silica remains, both fossil and extant, could be applied as bio-indicators for paleolimnology reconstruction of water ecosystems. The diatom

J. F. Ponce and M. Fernández, *Climatic and Environmental History of Isla de los Estados, Argentina*, SpringerBriefs in Earth System Sciences, DOI: 10.1007/978-94-007-4363-2_8, © The Author(s) 2014

study of Isla de los Estados presented here is the first one developed on the island. Although it was part of a doctoral thesis (Fernández 2013), some results of the diatom analysis have already been published by Unkel et al. (2010) and Fernández et al. (2012). Until today, diatom analyses of paleoarchives in southern Patagonia have not been extensive or systematic. Paleoenvironmental studies began with Frenguelli (1924) on a peat bog called La Misión (53° 30'S; 67° 50'W), located near the city of Río Grande (northern Tierra del Fuego) and continued with sediment analysis at Lago Fagnano, Tierra del Fuego (M. Espinosa, in Bujalesky et al. 1997; Recasens 2008). More recently, diatom studies from Puerto del Hambre in the Magellan Straits (McCulloch and Davies 2001), Las Cotorras peat bog in Tierra del Fuego (Borromei et al. 2010), and from other lake sediment sequences in southern Santa Cruz, such as Potrok Aike (Mayr et al. 2005; Wille et al. 2007; Fey et al. 2009) have been published.

8.2 Field and Laboratory Methods

The analyzed sub-samples were taken from two sediment cores (Laguna Cascada and Lago Galvarne Bog) located in the central part of Isla de los Estados (Fig. 8.1).

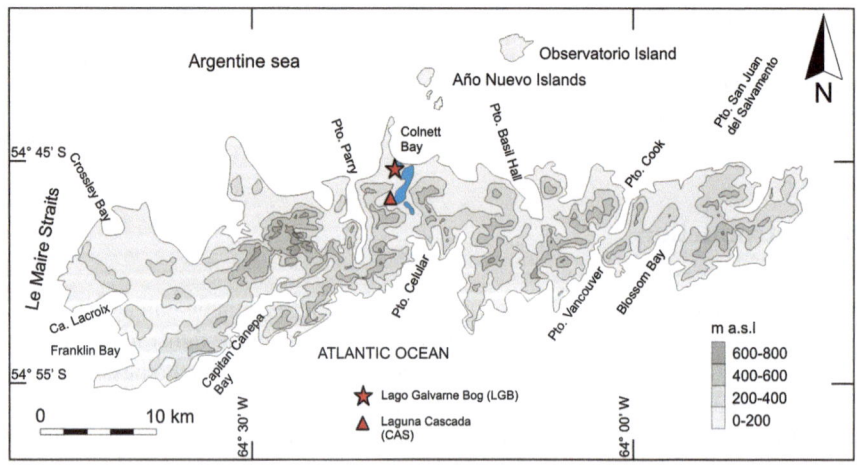

Fig. 8.1 Map of Isla de los Estados, showing the location of Laguna Cascada and Lago Galvarne Bog cores

8.2.1 *Laguna Cascada Core (54° 45'51"S, 64° 20'20.7"W, ca. 10 m a.s.l.)*

A 523 cm sediment core was retrieved with a Russian 7.5 cm-diameter corer and 1 m-long chamber (Fig. 8.2). Overlap between individual 1 m sections was at least 20 cm. Sub-samples for diatom analysis were taken at 2–3 cm intervals. In total 114 sub-samples were analyzed.

8.2.2 *Lago Galvarne Bog Core (54°44'16"S, 64° 19'37.9"W; 2 m a.s.l.)*

A 745 cm sediment core was retrieved with a Russian 5 cm-diameter corer and 1 m-long chamber (Fig. 8.3). Overlap between individual 1 m sections was at least 20 cm. A 65 sub-samples in the 232 a 448 cm section were analyzed.

8.3 Results

8.3.1 *Laguna Cascada*

The sediment sequence of Laguna Cascada (CAS) was divided in 20 stratigraphi-cal units. Diatom analysis was carried out between 528 and 163 cm deep and

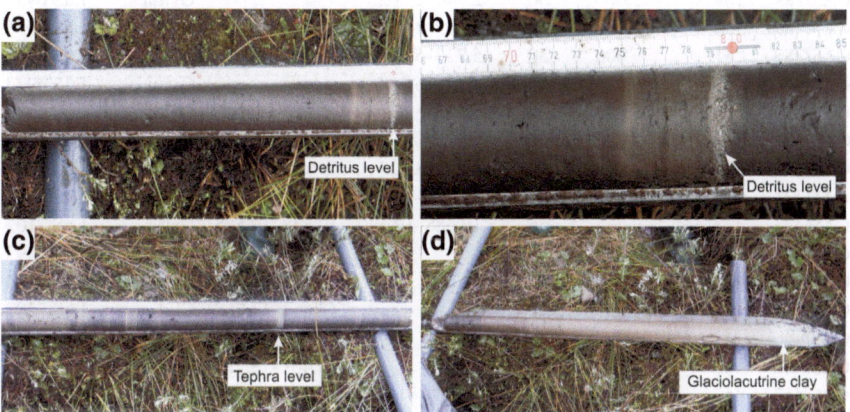

Fig. 8.2 Different sections of Laguna Cascada sediment core (Isla de los Estados). **a, b** Part of the core show a disconformity, in picture **b** the detritus level. **c** Different colors of gyttja. In one of the units there was a tephra layer. **d** deepest part of the core, characterized by a glacio-lacustrine clay

Fig. 8.3 Different sections of Lago Galvarne Bog (Isla de los Estados). **a** One of the section formed by gyttja. **b** Part of the core in which some sand layers are between organic gyttja layers

include from clay sediments to silt gyttja of variable colors. The chronology used here follows the one already published by Unkel et al. (2008, 2010) (Table 8.1).
A total number of 230 taxa were identified. The diatom taxa with a relative abundance higher than 3 % are shown in Fig. 8.4. The result of cluster analysis showed three zones with different sub-zones (Fig. 8.4). The *Aulacoseira* spp. group includes the following species: *A. alpigena*, *A. ambigua*, *A. distans*, *A. laevissima*, *A. subartica*, and *A. tethera*.

Table 8.1 Radiocarbon dates and calibrated ages from Laguna Cascada (CAS)

Sample N°	Depth (cm)	14C Age (BP)	1 σ error (BP)	SHCal 04a	IntCal 04a	Sample material	Lab. reference
CAS/126	126	1,290	50	1,256–1,083	–	Gyttja	LuS 6929
CAS/162	162	2,605	50	2,748–2,503	–	Gyttja	LuS 6506
CAS/189	189	2,600	50	2,746–2,502	–	Gyttja	LuS 6930
CAS/235	235	4,115	50	4,785–4,438	–	Gyttja	LuS 6931
CAS/281	281	5,000	50	5,720–5,609	–	Gyttja	LuS 6932
CAS/324	324	5,675	50	6,445–6,318	–	Gyttja	LuS 6933
CAS/346	346	5,920	50	6,775–6,571	–	Gyttja	LuS 6934
CAS/387	387	7,715	60	8,537–8,400	–	Gyttja	LuS 6935
CAS/391	391	7,775	60	8,548–8,429	–	Gyttja	LuS 6936
CAS/421	421	9,125	60	10,283–10,178	–	Gyttja	LuS 6937
CAS/423	423	9,435	55	10,688–10,520	10,728–10,585	Gyttja	LuS 7102
CAS/428	428	9,730	55	–	11,225–11,125	Clay-gyttja	LuS 7293
CAS/433	433	10,085	60	–	11,815–11,408	Clay-gyttja	LuS 7294
CAS/438	438	10,790	70	–	12,852–12,760	Clay-gyttja	LuS 7200
CAS/440	444	11,005	80	–	13,020–12,870	Clay-gyttja	LuS 6508
CAS/448	448	11,505	60	–	13,401–13,286	Clay-gyttja	LuS 7295
CAS/457	457	12,120	80	–	14,062–13,871	Clay-gyttja	LuS 6938
CAS/477	478	12,575	80	–	15,009–14,592	Clay-gyttja	LuS 6939
CAS/490	490	12,935	80	–	15,424–15,111	Gyttja-clay	LuS 6940
CAS/495	499	13,285	80	–	15,949–15,545	Gyttja-clay	LuS 6507

Modified from Unkel et al. (2008) and Unkel et al. (2010)

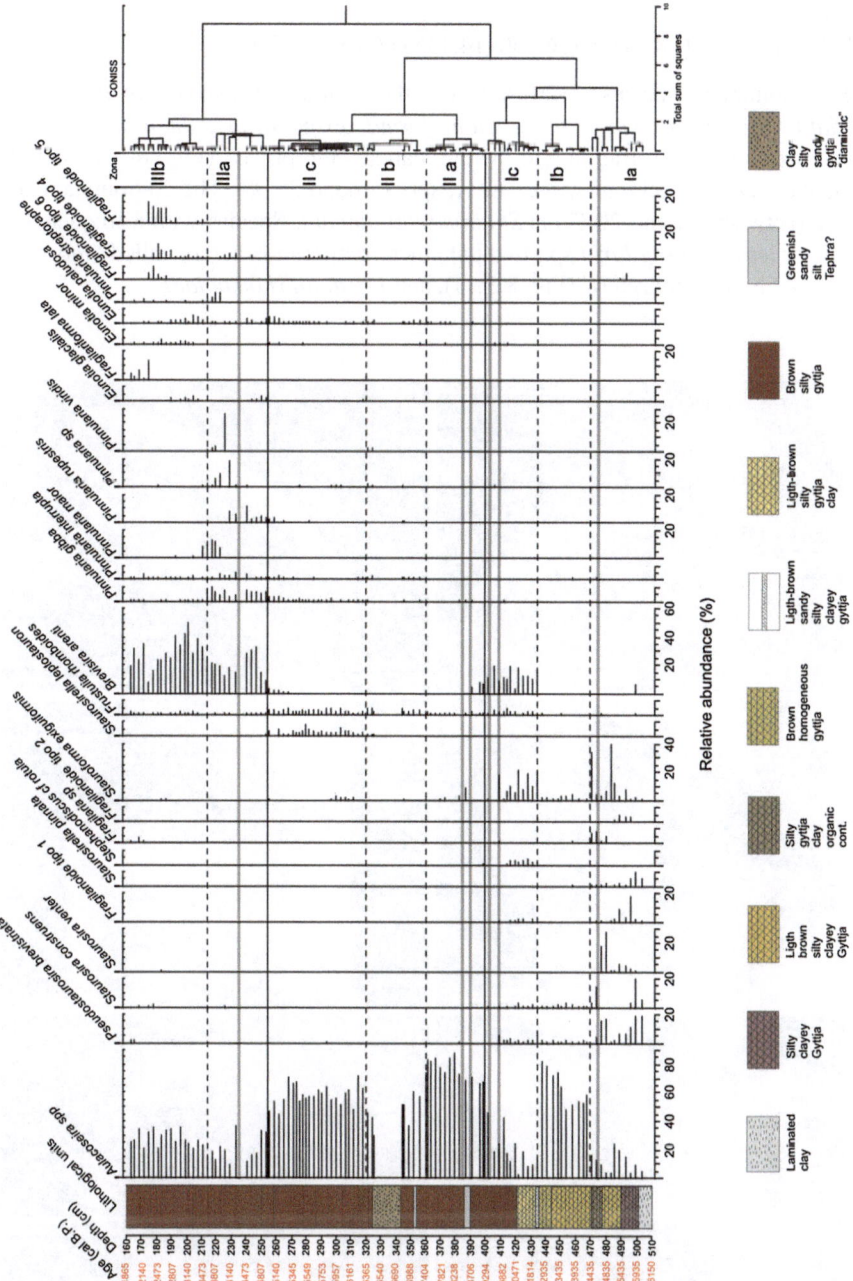

Fig. 8.4 Diatom assemblages from Laguna Cascada and the cluster analysis CONISS. The species shown are present in more than 3 % of relative abundance. There are three main zones. The *gray lines* show different thephra layers

Zone I

Sub-zone Ia: 510–470 cm (16.150–14.435 cal. years B.P.)

The dominant species (>20 %) are the benthic *Pseudostaurosira brevis-triata*, *Staurosira construens*, *Stauroforma exiguiformis*, *Staurosira venter,* and one unknown fragilarioid herein called "morph 1" (Fig. 8.5a), and the planktonic/tycoplanktonic *Aulacoseira* spp. (around 25 %). The co-dominant species (between 10 and 20 %) is *Staurosirella pinnata*. Species with a frequency of 5 % or less are *Fragilaria* sp., two unknown fragilarioids herein called "morph 2" (Fig. 8.5b) and "morph 6" (Fig. 8.5e, f), and *Frustulia rhomboides*.

Fig. 8.5 Identified species in Laguna Cascada (MEB). **a** Valvar view of *Fragilarioid morph 1*. **b** Valvar view of *Fragilarioide morph 2*. **c, d** *Aulacoseira alpigena*: valvar view (**c**), girdle view (**d**). **e, f** *Aulacoseira perglabra*: girdle view (**e**), valvar view (**f**). Scale: 5 μm

At 475 cm a tephra layer was recorded, probably corresponding to the first volcanic eruption of Mount Reclus (Southern Chile; 12.685 ± 260 ^{14}C years B.P.)
(McCulloch et al. 2005; Stern 2008; Unkel et al. 2008).

Sub-zone Ib: 470–435 cm (14.435–12.529 cal. years B.P.)

A notable increase of *Aulacoseira* spp. (80 %) and in parallel, a decrease in the
relative abundance (≤al 5 %) of *Pseudostaurosira brevistriata*, *Staurosira construens* and *Stauroforma exiguiformis* took place here.

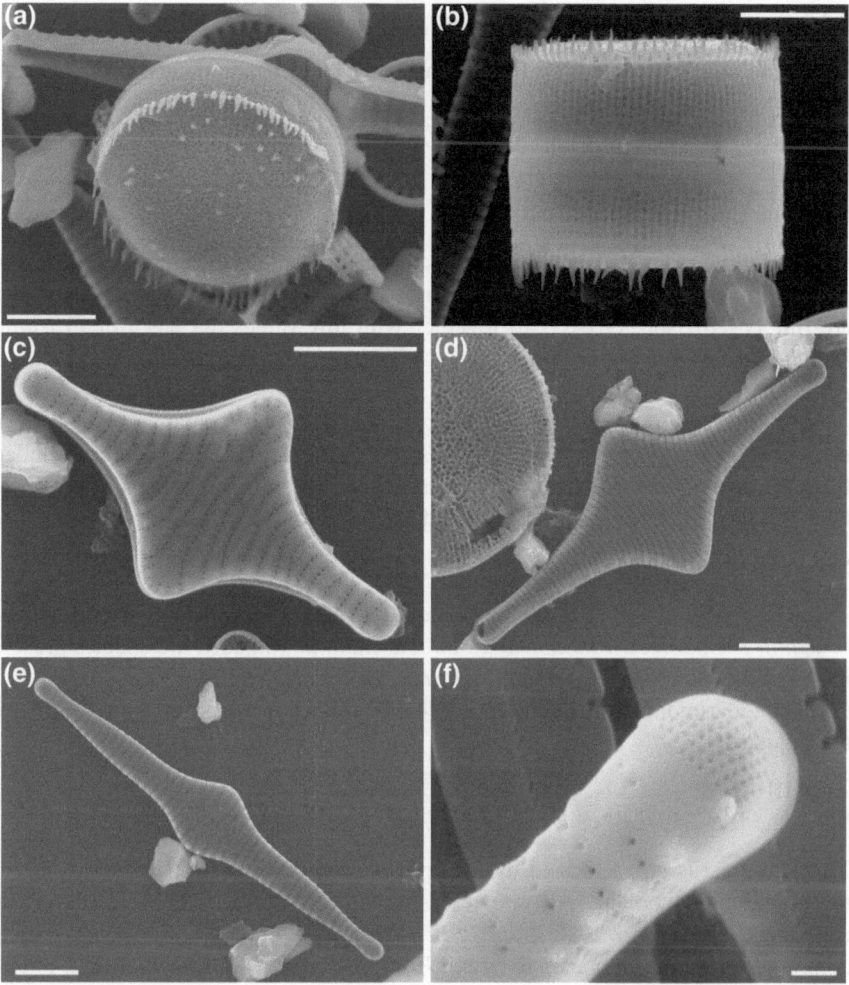

Fig. 8.6 Identified species in Laguna Cascada (MEB). **a, b** *Brevisira arentii*: valvar view (**a**),
girdle view of *Brevisira arentii* (**b**). **c** Valvar view of *Fragilarioide morph 4*. **d** Valvar view of
Fragilarioide morph 5. **e, f** *Fragilarioide morph 6:* valvar view (**e**), apical field detail (**f**). Scale:
5 μm

Sub-zone Ic: 435–400 cm (12.529–9,294 cal. years B.P.)

The *Aulacoseira* spp. frequency decreases in some moments reaching values around
5 %. The dominant species are *Brevisira arentii* (Fig. 8.6a, b) and *Stauroforma exiguiformis* (20–25 %). *Frustulia rhomboides* begins to increase its frequency at 420 cm.
Also, the planktonic species *Stephanodiscus* cf *rotula* is shows in the sub-zone and the
fragilarioid "*morph 1*" reappears. *Pseudostaurosira brevistriata and Staurosira construens* are present in a low frequency. The benthic species *Pinnularia gibba* and *Eunotia
minor* are identified in a very low frequency as well.

Zone II
Sub-zone IIa: 400–360 cm (9,294–7,404 cal. years B.P.)

This zone is characterized by the absence of *Brevisira arentii* and a remarkable
increase in *Aulacoseira* spp. These ones reach the highest frequency in the entire
profile (85 %). Also *Stauroforma exiguiformis, Frustulia rhomboides, Pinnularia
gibba* and *Eunotia paludosa* are registered in low frequencies.

Sub-zone IIb: 360–320 cm (7,404–6,365 cal. years B.P.)

This sub-zone is interrupted by an erosive discordance at 345 cm deep and the
presence of diamictic sediments of the unit 14. The main species in this sub-zone
are *Aulacoseira* spp. (75–85 %) together with low values (ca 5 %) of *Frustulia
rhomboides, Staurosirella leptostauron, Pinnularia interrupta* and *E. paludosa.*
All these species are acidophilous.

   Also, two levels in unit 14 are analyzed but these only contain a low quantity
of valves (only between 20 and 50 valves per slide), for which they were excluded
from the analysis. However, the diatom species are almost the same to those contained in the sediments from units 13 and 15.

Sub-zone IIc: 320–255 cm (6,365–4,973 cal. years B.P.).

Aulacoseira spp. dominates this sub-zone (up to 80 %). *Staurosirella leptostauron,
Frustulia rhomboides, Pinnularia gibba* and *Eunotia paludosa* are species present
in low frequency *Stauroforma exiguiformis* and *Brevisira arentii* reappears. Also,
a new fragilarioid herein called "morph 4" (Fig. 8.6c) appears for the first time in
the sequence.

Zone III
Sub-zone IIIa: 255–214 cm (4,973–3,607 cal. years B.P.)

This sub-zone is characterized by a significant decrease in relative abundance of
the planktonic/tycoplanktonic species (*Aulacoseira* spp.) and an increase in the
abundance of *Brevisira arentii* (ca 35 %). Also acidophilous and benthic species,
common in peatbogs, appear: *P. gibba, P. interrupta, P. maior, P. rupestris, P. vir-
idis,* and *P. streptoraphe.* At the beginning of the zone, low values of *Eunotia glacialis* are recorded. It is evident that *Aulacoseira* spp. reaches a peak in its relative
frequency immediately after the tephra deposition coming from the second eruption of Mt. Burney, called MB_2 (Stern 2008).

Sub-zone IIIb: 214–160 cm (3,607–1,830 cal. years B.P.)

Brevisira arentii reaches its maximum percentages (40–50 %) in this sub-zone, whereas the values of *Aulacoseira* spp. are between 30 and 40 %. Several species indicative of oligotrophic conditions are identified (for example, *Eunotia paludosa*, *E. minor*, *Pinnularia interrupta*). The fragilarioids herein called "morph 4," "morph 5," and "morph 6" (Fig. 8.6d–f) are recorded. The ecological information is unknown. To the top of the sequence *Fragilariforma lata* becomes important. In general, the species number associated to peat bog environments increases and dominates at the end of this sub-zone.

8.3.2 Lago Galvarne Bog

The sediment sequence from Lago Galvarne Bog (LGB) was divided in 12 stratigraphic units, but only five of them were studied to evaluate its diatom content. In this profile, 151 different taxa were identified. The diatom taxa with a relative frequency higher than 3 % are represented in Fig. 8.7. Some samples were excluded from the general statistical analysis because they did not reach the acceptable minimum number of valves. The cluster analysis result allows the division of the sequence in two zones. The first one presents two sub-zones. Five unidentified taxa of the genera *Cocconeis* were grouped in a preliminary order as *Cocconeis* spp. (Fig. 8.8c).

The chronology used here follows the one already published by Unkel et al. (2008, 2010) (Table 8.2).

Zone I
Sub-zone LGB Ia (448–404 cm; 7,430–6,608 cal. years B.P.)

Around the beginning of the sub-zone there are high values of marine epiphytic forms as *Cocconeis scutellum* (Fig. 8.8d, e) and *Opephora olsenii*, *C. disculus*, *C. placentula* (Fig. 8.8a), *Cocconeis* spp. and *Gyrosigma nodiferum*. Among those whose ecology was studied, all are epiphytic forms from brackish to fresh waters (Krammer and Lange-Bertalot 1986; Vos and de Wolf 1993). Also, the epiphytic *C. neodiminuta*, the tycoplanktonic *Staurosirella pinnata*, and *Stauroforma* cf *exiguiformis* are identified. In the first part of the sub-zone epiphytic and benthic species (*Eunotia* sp.) dominate. From the middle part to the end of the sub-zone, a planktonic species from brackish waters *Cyclotella meneghiniana* are present together with a high frequency of *Platessa oblongella*. Also, *Anamoeoneis brachysira* and *Stauroforma* cf *exiguiformis* are present. Lago Galvarne still seemed to have a slight marine water influence.

Sub-zone LGB Ib (404–290; 6,608–3,696 cal. years B.P.)

In this sub-zone there is a record of a high relative frequency of epiphytic and planktonic species: *Cocconeis* spp., *C. placentula*, *C. scutellum*, *C. neodiminuta*, *Gyrosigma nodiferum*, *Cyclotella meneghinina* and *Aulacoseira ambigua*. Other

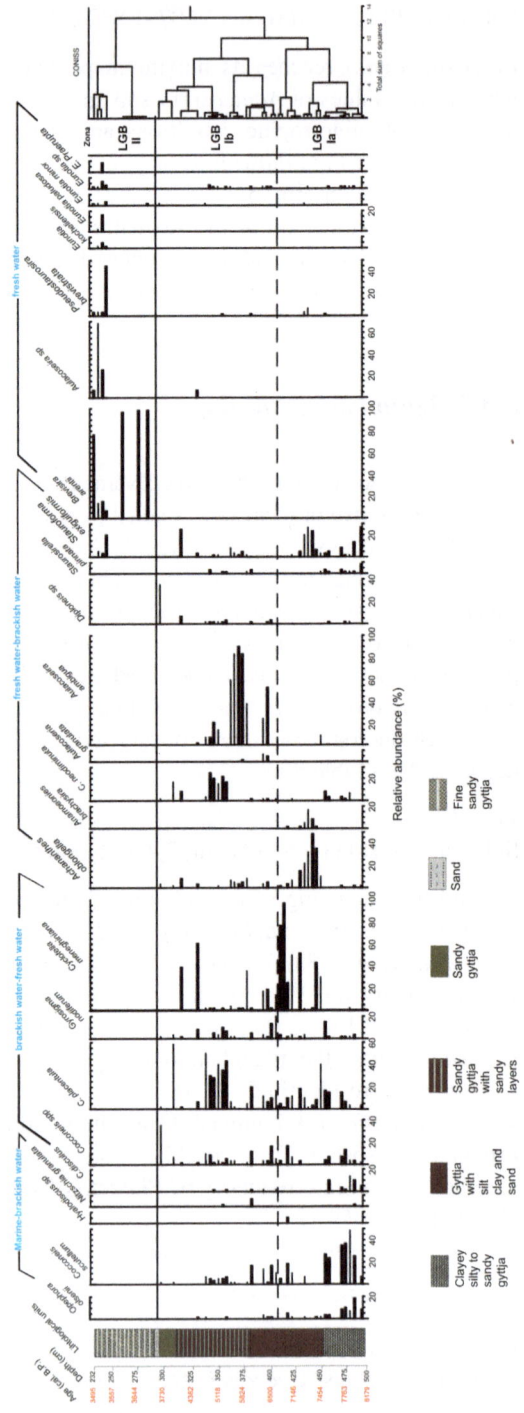

Fig. 8.7 Diatom assemblages from Lago Galvarne Bog and the cluster analysis CONISS. The species shown are present in more than 3 % of relative abundance. There are three main zones. The *gray lines* show different thephra layers

Fig. 8.8 Identified diatoms in Lago Galvarne Bog (MEB). **a** Valvar view of *Cocconeis placentula*. **b** Valvar view of *Diploneis* sp. **c** Valvar view of *Cocconeis* sp 2. **d** Valvar view of *Cocconeis scutellum*. **e** Valve view of the rapheless valve of *C. scutellum*. **f** Detail of the central area of *Gyrosigma nodiferum*. Scale: 5 μm

species with low frequencies are: *Opephora olsenii, Nitzchia granulata*, benthic, marine-brackish (Vos and de Wolf 1993), *C. disculus, Platessa oblongella, Diploneis* sp. (Fig. 8.8b), *Staurosirella pinnata, Stauroforma exiguiformis* and *Eunotia* sp. and also, some peaks of *Aulacoseira* sp., *Pseudostaurosira brevistriata* and *Eunotia minor*. This sub-zone is coincident with the stratigraphical units 6, 7, and 8 mainly (Fig. 8.6) formed by sand gyttja fine and coarse detritus intercalated.

Table 8.2 Radiocarbon dates from Lago Galvarne Bog (LGB) and calibrated ages

Sample N°	Depth (cm)	14C Age (BP)	1 σ error (BP)	SHCAl 04a	IntCal 04b	Sample material	Lab. reference
LGB/31	31	310	50	450–280	–	Peat	LuS 6791
LGB/125	130	1,665	65	1,570–1,410	–	Peat	LuS 6512
LGB/210	216	3,245	50	3,460–3,360	–	Coarse detritus	LuS 6792
LGB/300	303	3,460	60	3,820–3,560	–	Coarse detritus	LuS 6509
LGB/375	365	4,870	50	5,610–5,470	–	Gyttja	LuS 6793
LGB/433	423	6,210	60	7,160–6,970	–	Gyttja	LuS 6794
LGB/500	490	7,050	65	7,930–7,750	–	Peat	LuS 6511
LGB/547	539	8,130	55	9,120–8,780	–	Peat	LuS 6795
LGB/609	610	10,005	60	11,600–11,240	–	Peat	LuS 6796
LGB/632	631	10,410	65	–	12,680–12,500	Peat	LuS 7199
LGB/662	661	12,380	70	–	14,590–14,160	Peat	LuS 6797
LGB/700	699	13,030	80	–	15,560–15,210	Peat	LuS 6798
LGB/749	744	13,515	60	–	16,260–15,860	Peat	LuS 6510

Modified from Unkel et al. (2008, 2010)

Zone II
LGB II (290–232 cm; 3,696–3,495 cal. years B.P.)

The dominant species is *Brevisira arentii*, which is common in acid and dystrophic to mesotrophic lakes (Krammer and Lange-Bertalot 1991). Also, *Stauroforma exiguiformis*, *Aulacoseira* sp., *Pseudostaurosira brevistriata* and benthic forms: *E. kocheliensis*, *E. paludosa*, *E. minor* and *E. praerupta* are identified. These taxa are representative of diverse environments such as fresh waters as well as brackish waters.

8.4 Interpretation

8.4.1 Laguna Cascada Profile

The beginning of the sedimentological sequence is composed of clays of a possibly glaciolacustrine origin (Fig. 8.9a) that were deposited during the deglaciation of the cirques that surrounded the catchment area of the lake (Unkel et al. 2008). After 16,000 cal. years B.P., the sediments increasingly display their organic character, which is corroborated with gradual increases in both the values of TOC ("total organic carbon") and the ratio of C/N (Carbon vs. Nitrogen; Unkel et al. 2008) (Fig. 8.3). Between 16,000 cal. years B.P. and 15,200 cal. years B.P., the benthic species are predominant, especially the fragilarioids (Fig. 8.9a). These species can be found in shallow and alkaline waters, peripheral areas to deep lakes and oligotrophic environments (Douglas and Smol 2010). They have the ability to tolerate a wide spectrum of environments and they are found in almost all aquatic systems (Wilson et al. 1997); they can also survive under extreme environmental

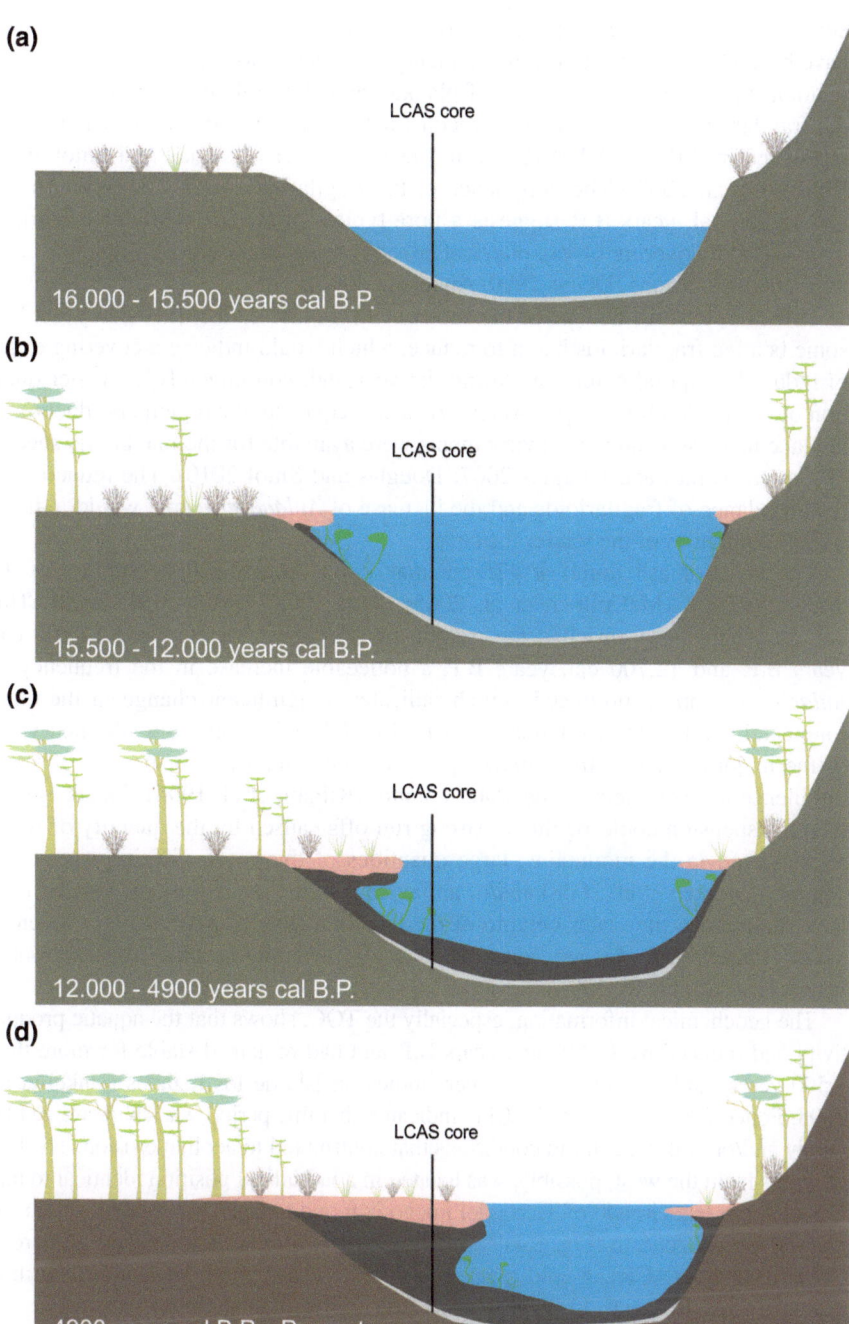

Fig. 8.9 Diagram of Laguna Cascada evolution. **a** Lake without vegetation. **b** The lake level increased and the aquatic vegetation began to grow. **c** The peat vegetation started to colonize the lake margin and the open forest reached the area. **d** Development of a peatbog over the lake, together with the presence of a close forest

conditions such as lakes or ponds in polar regions (Stoermer 1993). Likewise, they
have been recorded in proglacial sediments in alpine regions (Haworth 1988). The
relationship between the quantity of planktonic and periphytic diatoms, including
the fragilarioids, has been used to reconstruct the duration of the seasonal covering of ice in different lakes (Smol and Douglas 2007; Douglas and Smol 2010;
Rühland et al. 2008). The dominance of the fragilarioid species between 16,000
and 15,200 cal. years B.P. suggests a flora typical of environments characterized
by a seasonal covering of ice, physical erosion of the environment and harsh environmental conditions (Denys 1990; Anderson 2000).

However, towards the end of the sub-zone Ia (Fig. 8.4), at 14,800 cal. years B.P.
some benthic fragilarioids begin to reduce, which would indicate a covering of ice
of reduced temporal extension during the year and, consequently, a greater duration of the availability of open waters. As a consequence, the organic production of
the lake increased and new environments were available for the planktonic species
of diatoms (Smol and Douglas 2007; Douglas and Smol 2010). The reduction in
the abundance of fragilarioids and the increase of Aulacoseira spp. would indicate
a shorter duration of the winter ice cover.

The layer of ash found at 475 cm may come from the first eruption of the
Reclus volcano (McCulloch et al. 2005; Stern 2008; Unkel et al. 2008). This
deposit would have enriched the trophic state of the lake. Between 14,300 cal.
years B.P. and 12,700 cal. years B.P. a noticeable increase in the frequency of
Aulacoseira spp. is produced, which indicates a significant change in the environmental conditions of Laguna Cascada (Fig. 8.9b). The strongly silicified valve
of the majority of the Aulacoseira spp. need moments of re-suspension through
turbulence to keep them in the water column (Kilham et al. 1996). The causes of
the re-suspension could be due to strong run-offs caused by the quantity of water
accumulated in the mountains, large quantities of thaw water during spring and
summer (Sterken et al. 2008) and/or an increase in the wind velocity, which could
have maintained the water column of the lake in a state of permanent turbulence.
Many Aulacoseira spp. are common in acidic and oligotrophic to mesotrophic
waters (van Dam et al. 1994).

The geochemical information, especially the TOC, shows that the aquatic productivity had reduced by 14,500 cal. years B.P. and had remained stable for more than
1,500 years (Unkel et al. 2008). Other studies on Isla de los Estados (Unkel et al.
2008; Ponce 2009; Ponce et al. 2011) indicated that this period was characterized by
windy and/or arid to semiarid conditions that contributed to aeolian erosion. The belt
of winds from the west, possibly, was located in a latitudinal position identical to that
in which the Isla Grande of Tierra del Fuego is found today. This coincides with the
diatomological zone Ib (Fig. 8.4) and corresponds with the period called "Antarctic
Cold Reversal" (ACR), a cold phase registered for the Southern Hemisphere that
extended from 14,500 to 12,800 cal. years B.P. (Hubbard et al. 2005).

The gyttjia clayey silt that appears at 435 cm (12,700 cal. years B.P.) coincides with a change in the assemblage of diatoms. Here the sub-zone Ic begins
(Fig. 8.4) up to 400 cm (9,200 cal. years B.P.). The species that dominate this sector are Brevisira arentii, Stauroforma exiguiformis and various Aulacoseira spp.

The first is found preferentially in shallow lakes with muddy waters (Flower 2005) and associated with high concentrations of dissolved organic carbon (DOC, Köster and Pienitz 2006) in the water column. The presence of these diatoms suggests the development of a new phase in Laguna Cascada. The sediments and the geochemical analyses indicate an increase in the productivity and warm and quite arid conditions (Unkel et al. 2008). These warmer conditions would have enabled the growth of aquatic vegetation around the lake, so providing habitats and substrates for the growth of the diatoms (Fig. 8.9c) (Douglas and Smol 1995).

The beginning of zone II (9,200 cal. years B.P.) is characterized by the deposition of a thick layer of ash, probably corresponding to the first eruption of the Hudson volcano. The addition of new minerals, especially silica, contributed to modifying the trophic state of the lake. This would have favoured a large expansion of the genus *Aulacoseira*. The great abundance of these species suggests high lake levels and a very good mixing in the water column. Also, the very good mixing could suggest a period characterized by a high frequency and intensity of the SHW. The erosive discontinuity between 325 and 345 cm corresponds to an intense episode of surface run-off, possibly alluvial in type. The samples analysed, corresponding to this discontinuity, contain few diatom valves, which appear very fragmented. The amount of bromine (Br) in the sediments of Laguna Cascada increased at 8,500 cal. years B.P. (Fig. 8.4) and remained high until 4,500 cal. years B.P. (Unkel et al. 2010). This element comes from the marine salts. The increase of the Br is possibly related to an invasion of marine water in Lago Lovisato, located 300 m, in a NE direction from Laguna Cascada, during the transgressive event of the Middle Holocene. The Br would have arrived in the basin of Laguna Cascada by way of marine aerosols coming from Lago Lovisato during the interval of time during which the marine incursion lasted in that lake.

After 4,900 cal. years B.P. and until 3,500 cal. years B.P. the noticeable reduction in the group of *Aulacoseria* spp. and the reappearance in the record of *Brevisira arentii* together with benthic species related to peat bog environments (Figs. 8.4 and 8.9d) suggest very wet conditions and a large development of macrophytes in the margins of the lake. The vegetation began to grow progressively from the margins advancing towards the centre of the lake. This would indicate the sudden passage from a purely lacustrine environment to a peat bog in this sector of the lake (Fig. 8.9d). The hydrological variability is reflected in the change of relative abundance of benthic species to planktonic or vice versa (Gaiser and Rülhand 2010). This leads us to interpret the change of the assemblages of diatoms as a result of the growth of vegetation belonging to peat bogs in Laguna Cascada. Also, it could be linked to differences in the bathymetry of the lake, where the shallower zones may have resulted in peat bogs and the deeper zones may have remained as a lake.

It is important to highlight the deposition around 235 cm of an ash level, probably corresponding to the second eruption of Mount Burney (Unkel et al. 2010). Between 3,500 and 1,865 cal. years B.P., *Brevisira arentii* attains a peak of 50 % and later begins to reduce with a corresponding increase in the benthonic species and those belonging to the peat bog, which could indicate that the environment was reaching the equilibrium.

8.4.2 Lago Galvarne Peat Bog: Evidence of Sea Level Increase and Paleoenvironmental Development

The Lago Galvarne peat bog has marine deposits of some 4.5–5 m in thickness. The sediments of Lago Galvarne indicate the development of a humidified peat around 6 m (unit 3), later evolving into a poorly humidified peat (unit 4a) to conclude, before the marine influence, in a peat rich in minerals (unit 4b). That is to say, that until before 8,100 cal. years B.P. a continental type environment dominated in the basin, associated with the development of the peat bog. After this, the diatom studies suggest a strong influence of marine waters in the basin, as shown by the predominance of marine species such as *Ophephora olsenii, Cocconeis scutellum,* and *Nitzschia granulata* accompanied by salty water species like *Cocconeis placentula, Gyrosigma nodiferum,* and *Cyclotella meneghiniana* between 8,000 and 7,400 cal. years B.P. (Fig. 8.7). Also, the geochemical analyses indicate a gradual increase in the values of Bromine (Br) after 8,500 cal. years B.P., which could be explained by the increase in the influence of sea water in the lake. This increment of the main influence in the basin would indicate the presence of a transgressive–regressive event, possibly in correlation with the transgression of the Middle Holocene, an event recorded on a regional and global scale. Between 7,400 and 3,700 cal. years B.P. a lagoon, or maybe a fjord type of environment had developed, evidence for which is seen in the high frequency of brackish species with some fresh water and epiphytes, such as: *Cocconeis placentula, Gyrosigma nodiferum, Cyclotella meneghiniana, Platessa oblongella, Stauroforma exiguiformis* (Fig. 8.7). After 3,700 cal. years B.P. the basin no longer had marine influence or salty waters. This can be observed in the stratigraphic units 8 and 9, which have a high organic content (40–60 %), high values of C/N; and low values in concentration of Br (bromine) and Fe (iron) ions and magnetic susceptibility (Fig. 8.5) (Unkel et al. 2010). The diatoms identified in these units are exclusively from fresh water. The dominant species is *Brevisira arentii* (also recorded for Laguna Cascada), accompanied by *Pseudostaurosira brevistriata, Aulacoseira* sp., *Eunotia kocheliensis, Eunotia paludosa, Eunotia minor, Eunotia praerupta,* and *Eunotia* sp. with variable frequencies. The marine influence would have ended due to a new global eustatic descent, accompanied by the development of an extensive line of beach or littoral band, which separated the current Lago Galvarne from the sea. The development of these littoral bands can occur due to exceptional meteorological events such as storms or "ice push"; by tidal waves or by repetition of tectonic events (Gordillo et al. 1992). For Isla de los Estados the last two of these factors are considered the most probable (Ponce et al. 2009). The most modern units were not subject to diatomological analysis but in accordance with the sedimentological studies, the basin of Lago Galvarne would have evolved from lightly lacustrine conditions towards other conditions strongly influenced by the development of peat bog vegetation.

References

Anderson NJ (2000) Diatoms, temperature and climatic change. Eur J Phycol 35:307–314

Björck S, Rundgren M, Ljung K, Unkel I, Wallin A (2012) Multy-proxy analysis of a peat bog on Isla de los Estados, easternmost Tierra del Fuego; a unique record of the variable Southern Hemisphere Westerlies since the last deglaciation. Quatern Sci Rev 42:1–14

Borromei AM, Coronato A, Franzén LG, Ponce JF, López Sáez JA, Maidana N, Rabassa J, Candel MS (2010) Holocene paleoenvironments in subantarctic high Andean valleys (Las Cotorras mire, Tierra del Fuego, Argentina). Palaeogeography, Palaeoclimatology, Palaeoecology 286:1–16

Bujalesky GG, Heusser CJ, Coronato A, Roig C, Rabassa J (1997) Pleistocene glaciolacustrine sedimentation at Lago Fagnano, Andes of Tierra del Fuego, southernmost South America. Quaternary Sci Rev 16:767–778

Denys L (1990) *Fragilaria* blooms in the Holocene of the western coastal plain of Belgia. In: Simola H (ed) Proceedings of the X international diatom symposium. Koeltz Scientific Books, Finland, pp 397–406

Douglas MSV, Smol JP (1995) Periphytic diatom assemblages from high Arctic ponds. J Phycol 31:60–69

Douglas MSV, Smol JP (2010) Freshwater diatoms as indicators of environmental change in the High Artic. In: Smol JP, Stoermer EF (eds) The diatoms: applications for the environmental and earth sciences, 2nd edn. Cambridge University Press, Cambridge, pp 249–266

Fernández M (2013) Los paleoambientes de Patagonia Meridional, Tierra del Fuego e Isla de los Estados en los tiempos de las primeras ocupaciones humanas. Estudio basado en el análisis de diatomeas. Unpublished doctoral thesis, Universidad Nacional de La Plata, Facultad de Ciencias Naturales y Museo, La Plata

Fernández M, Maidana N, Rabassa J (2012) Palaeoenvironmental conditions during the Middle Holocene at Isla de los Estados (Staaten Island, Tierra del Fuego, 54° S, Argentina) and their influence on the possibilities for human exploration. Quatern Int 256:78–87

Fey M, Korr C, Maidana NI, Carrevedo ML, Corbella H, Dietrich S, Haberzettl T, Kuhn G, Lücke A, Mayr C (2009) Palaeoenvironmental changes during the last 1600 years inferred from the sediment record of a cirque lake in southern Patagonia (Laguna Las Vizcachas, Argentina). Palaeogeography, Palaeoclimatology, Palaeoecology 281:363–375

Flower R (2005) A taxonomic and ecological study of diatoms from freshwater habitats in the Falkland Islands, South Atlantic. Diatom Res 20:23–96

Frenguelli J (1924) Diatomeas de la Tierra del Fuego. Anales de la Sociedad Científica Argentina. 96:225–263; 97:87–118, 231–266; 98:63

Gaiser E, Rülhand K (2010) Diatoms as indicators of environmental change in wetlands and peatlands. In: Smol JP, Stoermer EF (eds) The diatoms: applications for the environmental and earth sciences, vol 25. Cambridge University Press, Cambridge, pp 473–496

Gordillo S, Bujalesky G, Pirazzoli PA, Rabassa J, Saliège JF (1992) Holocene raised beaches along the northern coast of the Beagle Channel, Tierra del Fuego, Argentina. Palaeogeogr Palaeoclimatol Palaeoecol 99:41–54

Haworth E (1988) Distribution of diatom taxa of the old genus *Melosira* (now mainly *Aulacoseira*) in Cumbrian waters. In: Round FE (ed) Algae and the aquatic environment. Biopress Ltd, Bristol, pp 139–167

Hubbard A, Hein AS, Kaplan MR, Hulton NRJ, Glasser N (2005) A modeling reconstruction of the last glacial maximum ice sheet and its deglaciation in the vicinity of the Northern Patagonian Icefield, South America. Geogr Ann A 87:375–391

Kilham SS, Theriot EC, Fritz SC (1996) Linking planktonic diatoms and climate change in the large lakes of the Yellowstone ecosystem using resource theory. Limnol Oceanogr 41:1052–1062

Kilroy C, Sabbe K, Bergey EA, Vyverman W, Lowe R (2003) New species of Fragilariforma (Bacillariophyceae) from New Zealand and Australia. J Bot 41:535–554

Köster D, Pienitz R (2006) Late Holocene environmental history of two New England pond: natural dynamics versus human impacts. Holocene 16:519–532

Krammer K, Lange-Bertalot H (1986) Bacillariophyceae. 1. Teil: Naviculariaceae. Süsswasser-flora von Mitteleuropa. Gustav Fisher Verlag, New York, p 876

Krammer K, Lange- Bertalot H (1991) Bacillariophyceae. 3 Teil: Centrales, Fragilariaceae, Eunotiaceae. Süssewasser-flora von Mitteleuropa. Spektrum Akademischer Verlag Heidelberg, Berlin, p 598

Mayr C, Fey M, Haberzettl T, Janssen S, Lücke A, Maidana NI, Ohlendorf C, Schäbitz F, Schleser GH, Struck U, Wille M, Zolitschka B (2005) Palaeoenvironmental changes in southern Patagonia during the last millennium recorded in lake sediments from Laguna Azul (Argentina). Palaeogeography, Palaeoclimatology, Palaeoecology 228:203–227

McCulloch R, Davies S (2001) Lateglacial and Holocene palaeoenvironmental change in the central Strait of Magellan, Southern Patagonia 173(3–4):143–173

McCulloch RD, Bentley MJ, Tipping RM, Clapperton CM (2005) Evidence for late-glacial ice dammed lakes in the central strait of Magellan and Bahía Inútil, southernmost South America. Geogr Ann A 87:335–362

Ponce JF (2009) Palinología y geomorfología del Cenozoico tardío de la Isla de los Estados. Unpublished doctoral thesis, Universidad Nacional del Sur. Bahía Blanca, Argentina

Ponce JF, Rabassa J, Martínez O (2009) Morfometría y génesis de los fiordos de la Isla de los Estados. Revista de la Asociación Geológica Argentina 65:638–647

Ponce JF, Borromei AM, Rabassa J, Martínez O (2011) Late quaternary palaeoenvironmental change in western Staaten Island (54°.5° S; 64° W), Fuegian Archipelago. Quatern Int 233:89–100

Recasens C (2008) Lago Fagnano, Tierra de Fuego: a multiproxy environmental record in southernmost Patagonia for the last ca. 200 years. Genève, Risques Gèologiques et Environnement. Falcuté des Sciences, Universitè de Genève

Rühland K, Paterson AM, Smol JP (2008) Hemispheric-scale patterns of climate induced shifts in planktonic diatoms from North American and European lakes. Glob Change Biol 14:2740–2745

Smol JP, Douglas MSV (2007) From controversy to consensus: making the case for recent climate change in the Arctic using lake sediments. Front Ecol Environ 5:466–474

Sterken M, Verleyen E, Sabbe K, Terryn G, Charlet F, Bertrand S, Boës X, Fagel N, De Batist M, Vyverman W (2008) Late quaternary climatic changes in southern Chile, as recorded in a diatom sequence of Lago Puyehue (40° 40′ S). J Paleolimnol 39:219–235

Stern C (2008) Holocene tephrochronology record of large explosive eruptions in the southernmost Patagonian Andes. Bull Volcanol 70:435–454

Stoermer EF (1993) Evaluating diatom succession: some pecularities of Great Lakes case. J Paleolimnol 8:71–83

Unkel I, Björck S, Wohlfarth B (2008) Deglacial environmental changes on Isla de los Estados (54.4° S), southeastern Tierra del Fuego. Quatern Sci Rev 27:1541–1554

Unkel I, Fernández M, Björck S, Kjung K, Wolfarth B (2010) Records of environmental changes during the Holocene from Isla de los Estados (54.4° S), southeastern Tierra del Fuego. Glob Planet Change 74:99–113

Van Dam H, Meriens A, Sinkeldam J (1994) A code checklist and ecological indicator values of freshwater diatoms from the Netherlands. Neth J Aquat Ecol 28:117–133

Vos CP, de Wolf H (1993) Diatoms as a tool for reconstructing sedimentary environments in coastal wetlands; metodological aspects. Hydrobiologia 269(270):285–296

Wille M, Maidana NI, Schäbitz F, Fey M, Haberzettl T, Janssen S, Lücke A, Mayr C, Ohlendorf C, Schleser GH (2007) Vegetation and climate dynamics in southern South America: the microfossil record of Laguna Potrok Aike, Santa Cruz, Argentina. Rev Palaeobot Palyno 146:234–246

Wilson SE, Smol JP, Sauchyn DJ (1997) A Holocene paleosalinity diatom record from southwestern Saskatchewan, Canada. Harris Lake revisited. J Paleolimnol 17:23–31

Chapter 9
Reconstruction of Paleoenvironmental Conditions During Late Glacial and Holocene Times in Isla de los Estados and Their Correlation with the Beagle Channel and Southern Patagonia

Abstract Late Glacial-Holocene environmental conditions were interpreted in Isla de los Estados from palynological and diatomological analysis and linked also with geoquimical studies. The deglaciation started after 16,000 cal. years B.P. followed by a rapid glacial retreat under gradually warmer conditions, with alternation of drier and wetter periods until ca. 12,800 cal. years B.P. Between ca. 12,600 and 10,300 cal. years B.P. the plant communities indicated moderately cold to mild and arid climatic conditions. During the Early Holocene the expansion of Nothofagus forests is interpreted as a signal of increasing temperature and precipitation in spite of the humidity levels being then lower than today. Around 8,000–6,000 cal. years B.P. the marine diatom assemblages and geoquimical analysis indicates the Middle Holocene marine transgression. An abrupt rise of arboreal taxa occurred between ca. 8,300 and 5,500 cal. yr BP related to increased effective precipitation that culminated with the establishment and persistence of closed-canopy forest communities of Sub-Antarctic Evergreen Forest. The diatoms assemblages suggest wetter conditions, and a greater development of aquatic vegetation from ca. 4,800 cal. years B.P. From ca. 2,700 cal. years B.P. an important vegetation change in the area can be seen, probably as a consequence of warmer and drier conditions.

Keywords Pollen • Diatoms • Paleoenvironment • Fuegian archipielago • Late glacial • Holocene

9.1 Late Glacial-Holocene (18,000–11,500 cal. years B.P.)

The Late Glacial is defined in the southern end of South America as the period between the beginning of the retreat of the ice from the moraines generated during the Last Glacial Maximum and the beginning of the Holocene, during which the global climatic conditions underwent significant and recurrent changes (Rabassa et al. 1992; Coronato et al. 1999). The Late Glacial-Holocene boundary has been conventionally established at 10,000 ^{14}C years B.P. (ca. 11,500 cal. years B.P.) (Rabassa et al. 2000).

J. F. Ponce and M. Fernández, *Climatic and Environmental History of Isla de los Estados,*   105
Argentina, SpringerBriefs in Earth System Sciences, DOI: 10.1007/978-94-007-4363-2_9,
© The Author(s) 2014

In the Beagle Channel, the recession of the Beagle Paleoglacier from its most extreme zone at Punta Moat (Fig. 9.1a) corresponding to the system of terminal moraines of the Last Glaciation (Moat Glaciation, Rabassa et al. 2000), began before 14,640 ^{14}C years B.P. (ca. 17,963 cal. years B.P.), which is indicated by the radiocarbon date obtained from the base of the Puerto Harberton peat bog, some 40 km southwest from Punta Moat (Heusser 1989a; Rabassa et al. 1990).

With the beginning of the deglaciation, the temperatures rose and the climatic conditions were more benign. In agreement with Pendall et al. (2001) who studied stable isotopes of hydrogen in mosses from the Puerto Harberton peat bog (Fig. 9.1a), the temperature would have increased dramatically from 0 °C, at 16,200 cal. years B.P. to 12 °C at 15,000 cal. years B.P. accompanied by conditions of lower effective humidity. Hall et al. (2013) presented data from Cordillera Darwin that showed rapid glacier recession in southern South America between 18,000 and 14,600 cal. years B.P.

In accordance with the fossil pollen records coming from the peat bogs located along the Beagle Channel (Heusser 1989a, 1998, 2003) from ca. 17,700 cal. years B.P. (14,640 ^{14}C years B.P.) the development of an impoverished postglacial vegetation is recorded, characterized by communities of scrub and low bushes, grasses and marsh taxa, with rare trees, belonging to steppe/tundra environments. Comparable vegetal communities are currently observed in the Patagonian steppe in northern Tierra del Fuego and along the Atlantic coast where the mean annual precipitation is less than 300 mm (Pisano 1977).

The low frequency of *Nothofagus* pollen recorded in all the fossil pollen profiles throughout the area of the Beagle Channel suggest the presence of small Sub-Antarctic woods coming from local glacial refuges, probably located very close to or in contact with the edges of the glacial ice along the channel and/or local periglacial environments (Premoli et al. 2010). The arboreal pollen information show highly variable records related with fluctuating conditions in the temperature environment. Two climatic deteriorations would have interrupted the progressive and slow expansion of the *Nothofagus* forest. One, around 13,000 ^{14}C years B.P. (15,300 cal. years B.P.), the "Antarctic Cold Reversal" (ACR) and the other, between 11,000 and 10,100–10,000 ^{14}C years B.P. (ca. 13,000 and 11,500 cal. years B.P.), the equivalent to the "Younger Dryas" (YD) of the Northern Hemisphere (11,000–10,000 14C years B.P.), with a noted absence of *Nothofagus* pollen at 10,200 ^{14}C years B.P. (ca. 12,000 cal. years B.P.) (Heusser and Rabassa 1987; Heusser 1998). A summer temperature for the latter event being estimated to be <3 °C lower than the current temperature at the location of the present city of Ushuaia and a drop in the precipitations in the order of 200 mm annually (Heusser 1998). This last colder Late Glacial episode, in the order of a millennium, was also identified in the pollen records of the inner Fuegian valleys (Borromei et al. 2007). For Sugden et al. (2005), the ACR affected the entire Patagonian region, being detectable in the area of San Carlos de Bariloche and in the Lake District of Chile, where another cold episode is recorded with high temporal resolution called the "Huelmo/Mascardi Cold Reversal" (Hajdas et al. 2003).

From the paleoecological information, paleoenvironmental conditions are inferred for the Late Glacial, that, although warmer, are mainly colder and drier

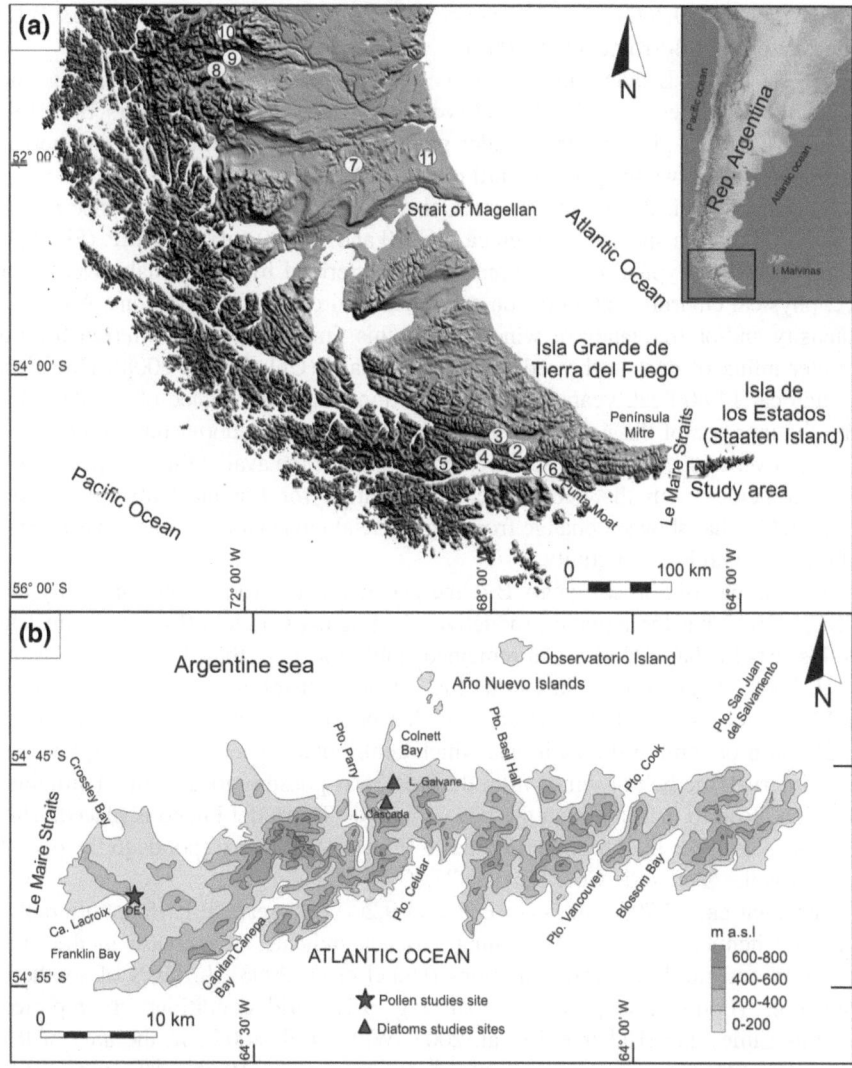

Fig. 9.1 Location map. **a** Southern Patagonia illustrating the location of the sites mentioned in the text: *1* Puerto Harberton (Heusser 1998); *2* Las Cotorras (Borromei et al. 2010); *3* Lago Fagnano (Borromei ct al. 2013); *4* Vallé de Andorra (Mauquoy et al. 2004); *5* Lago Roca-Bahía Lapataia; *6* Río Varela; *7* Laguna Potrok Aike (Wille et al. 2007); *8* Lago Guanaco (Moreno et al. 2009); *9* Vega Ñandú (Villa-Martínez and Moreno 2007); *10* Cerro Frías (Tonello et al. 2009); *11* Laguna Azul (Wille et al. 2007). **b** Isla de los Estados showing the location of pollen (Ponce et al. 2011) and diatoms (Fernández 2013) studied sites. Modified from Ponce et al. (2011)

than in the present times due to the local presence of the Beagle Channel pale-oglacier (Heusser 1998), and similarly for the glaciers in the interior valleys, until approximately 10,300–9,300 ^{14}C years B.P. (ca. 12,000–10,600 cal. years B.P.) (Coronato1995; Borromei et al. 2007).

In Isla de los Estados, in accordance with the oldest radiocarbon dating, the deglaciation started after 16,000 cal. years B.P. (ca. 13,500 ^{14}C years B.P.) followed by a rapid glacial retreat under gradually warmer conditions, with alternation of drier and wetter periods until ca. 12,800 cal. years B.P. (10,800 ^{14}C years B.P.) (Unkel et al. 2008). In Laguna Galvarne (Fig. 9.1b), the dominance of the fragilaroid diatom species between ca. 16,000 and 15,200 cal. years B.P. (Fig. 8.4) suggests a flora typical of environments characterized by a seasonal covering of ice, physical environmental erosion and harsh environmental conditions. A greater intensity and/or frequency of winds during this time interval are inferred from a greater influx of dust in the sediments of this lake (Unkel et al. 2008). However, around ca. 14,800 cal. years B.P., the reduction in the abundance of fragilaroides and the increase of the *Aulacoseira* spp. would indicate a shorter duration of winter ice cover and, consequently, a greater duration of the availability of open water. This coincides with the geochemical information for Laguna Galvarne (Unkel et al. 2008) that shows a change from a proglacial environment to an environment of greater organic productivity.

Around ca. 14,500 cal. years B.P. the geochemical information, especially the TOC, shows that the aquatic productivity in Laguna Cascada (Fig. 9.1b) on Isla de los Estados had reduced and remained stable for more than 1500 years (Unkel et al. 2008). Other studies in Isla de los Estados (Unkel et al. 2008; Ponce et al. 2011; Björck et al. 2012) indicate that this period was characterized by windy and/or arid or semi-arid conditions, which contributed to aeolian erosion, For this time interval, the belt of winds from the west, was possibly located in a latitudinal position close to that in which the Isla Grande de Tierra del Fuego is placed. This coincides with the diatomological zone Ib (Fig. 8.4) and corresponds to the period known as the "Antarctic Cold Reversal" (ACR).

Between ca. 12,700 cal. years B.P. and 9,200 cal. years B.P. the stratigraphy and geochemical analyses of Laguna Cascada indicate an increase in the productivity and mild and arid conditions (Unkel et al. 2008). In lakes of southern Patagonia (Potrok Aike, Laguna Azul Fig. 9.1a) arid conditions are reported for this same time (Haberzettl et al. 2007; Wille et al. 2007). In the area of the Magellan Strait an intense arid phase culminated at 10,650 ^{14}C years B.P. (McCulloch and Davies 2001). This sign is probably related to a more polar position of the westerly winds belt, resulting in a weak influence in the climatic conditions on Isla de los Estados.

The pollen record at Franklin Bay (western Isla de los Estados, Fig. 9.1b) shows evidence of herbaceous and marsh vegetation, without trees and with scarce bushes and scrub between ca. 12,600 and 10,300 cal. years B.P. (10,679–9,174 ^{14}C years B.P.). These plant communities would have developed in locally wetter areas, non- sloping, with slow surface run-off and deficient drainage, affected by the melt waters coming from the melting of nearby glaciers, under moderately cold to mild and arid climatic conditions. The minimum values of the relative frequency and concentration of *Nothofagus* pollen recorded in this interval are adjusted to the regional pattern of Late Glacial environments throughout the area of the Beagle Channel (Heusser 2003), with extra-regional sources arising (between 2 and 200 km distant; sensu Prentice 1985) contributing tree pollen probably coming from the wooded glacial refuges located to the west in the lowlands along the Beagle Channel and/or a very low pollen production due to unfavourable environmental conditions for the development of trees.

On the Isla Grande de Tierra del Fuego, the process of deglaciation would also have culminated towards the end of the Late Glacial and the beginning of the Holocene, not only in the lowlands along the Beagle Channel, but also in the interior tributary valleys of the Fuegian Andes (Borromei et al. 2007; Borromei and Quattrocchio 2008).

9.2 Holocene (ca. 11,500 cal. years B.P.—Present)

The beginning of the Early Holocene is seen in several records from Antarctica (ice cores, marine cores, and geomorphological archives) which express the continuous deglaciation (Bentley et al. 2009) due to the significant increase in temperature. This is what is known as the "Antarctic Thermal Optimum".

During the Early Holocene (ca. 11,500–8,000 cal. years B.P.) the pollen spectra indicate the development of transitional plant communities of the forest/steppe ecotone, associated with an open landscape with groups of trees and bushes in the lowland along the Beagle Channel (Heusser et al. 1998). Meanwhile, in the interior valleys an open vegetation of grassland and bushes existed and ice occupied the higher altitude areas in the hanging valleys (Borromei et al. 2007). Currently, a similar pattern of vegetation is developed in the central sector of the island where the annual precipitation does not exceed 400 m and the summer temperatures oscillate between 11 and 12 °C (Pisano 1977). This evidence suggests an increase in the temperature and effective humidity, although the precipitations remained lower than current levels. This would have favoured the development of fires, as can be seen in the record of carbon particles in the pollen profiles of the Beagle Channel area (Heusser 2003).

In the eastern end of Isla de los Estados, during the Early Holocene (between ca. 10,300 and 8,300 cal. yr BP), *Nothofagus* forests apparently expanded slowly at first on the landscape from their westward locations and formed patches into the heath, scrub and grass communities as a signal of increasing temperature and precipitation in spite of the humidity levels being then lower than today. Similar climate conditions, documented in the northern coast of the island, in Laguna Cascada (Unkel et al. 2008), were correlated with the onset of the Antarctic Thermal Optimum (Bentley et al. 2009). There, the assemblages of diatoms suggest the development of a new phase in Laguna Cascada. Milder conditions would have allowed the growth of aquatic vegetation around the lake, providing new habitats and substrates for the growth of benthic diatoms. The geochemical analyses for this locality indicate warm conditions and the establishment of denser vegetation around 9,400 ^{14}C years B.P. (ca. 10,600 cal. years B.P.) (Unkel et al. 2008). These climatic conditions would have been maintained until ca. 6,500 cal. years B.P. In general, from ca. 9,200 cal. years B.P. until ca. 5,000 cal. years B.P., the lithostratigraphy, geochemical analysis (Unkel et al. 2010) and the present diatomological studies indicate strong windy conditions, increase in precipitation and increase in the surface run-off processes. To the north, the low values of reconstructed annual precipitation from Cerro Frías pollen record located in southwestern Argentine Patagonia (Fig. 9.1a), were related to a weakening and southward shift of the westerlies during the Early Holocene (Tonello et al. 2009). Compilation of paleoclimate proxies from several Southern Ocean sediment cores and Antarctic ice cores suggested that the surface water warming, sea ice retreat and southward migration of the westerlies was a general feature of the Southern Ocean during the Early-Holocene Climate Optimum (Divine et al. 2010).

On Isla Grande de Tierra del Fuego the climatic improvement in the Early Holocene is coincident with a transgressive event in the Beagle Channel, which took place around 8,000 ^{14}C years B.P. (Rabassa et al. 2000). The incoming sea water flooded the area of Lago Roca-Bahía Lapataia (western end of the Beagle Channel, Fig. 9.1b) generating a deep and wide fjord with intricate archipelagos (Gordillo et al. 1993). This transgressive–regressive event generated several levels of marine beaches. The curve of the relative sea level in the Beagle Channel shows a maximum between 6,000 and 5,000 ^{14}C years B.P. as a consequence of the combined action of glacioeustatic increase and seismotectonic activity (Isla et al. 1999). The littoral vegetation in the time of the marine incursion in the Beagle Channel was mainly arboreal suggesting higher water availability under the moderating effect of the sea. In the pollen records of marine levels dated between 8,240–7,260 and 5,800 ^{14}C years B.P. in the locality of Bahía Lapataia and between 6,240 and 6,060 ^{14}C years B.P. in the locality of Río Varela (Fig. 9.1a), a significant increase is observed of *Nothofagus* pollen, suggesting the development of a dense forest along the coastal areas, at a time when a pattern of more open ecotone forest/steppe vegetation extended on a regional level (Grill et al. 2002; Borromei and Quattrocchio 2007).

In Isla de los Estados, the geochemical analyses from Laguna Cascada showed that the Br (bromine) of the sediments of this lake has two peaks: around 8,000 cal. years B.P. and 6,000 cal. years B.P. This element comes from the marine salts and increases in periods of strong and recurring storms (Unkel et al. 2010). The presence of Br possibly indicates the marine transgression of the Middle Holocene on Isla de los Estados, via the flooding with sea water of Lago Lavisato (Fig 9.1b).

In Lago Galvarne (Isla de los Estados), according to the sedimentological analysis and the interpretation of geochemical studies of the peat bog, as previously said, it is estimated that sea level would have increased around 8,500 cal. years B.P. (Unkel et al. 2010). Evidence for this is also seen in the assemblages of marine diatoms formed by Opephora olsenii, Cocconeis scutellum, Hyalodiscus sp., together with a high frequency of species from environments of high conductivities, such as C. placentula, Gyrosigma nodiferum, Cyclotella meneghiniana and Platessa oblongella that were then dominant (Fig. 8.7).

At Bahía Franklin, an abrupt rise of arboreal taxa occurred between ca. 8,300 and 5,500 cal. yr BP, as indicated by high percentage and concentration of *Nothofagus* pollen values (zone IDE-3) (Figs. 7.5 and 7.6). The arboreal pollen show frequent, large-amplitude fluctuations that indicate high variability in forest cover near the studied site. Between ca. 8,300 and 6,700 cal. yr BP (subzone IDE-3a), the landscape displays the physiognomy of a closed *Nothofagus* forest interspersed with dwarf shrub heath (*Empetrum*/Ericaceae type, *Myrteola nummularia*), scrub (Asteraceae subf. Asteroideae) and herb (*Gunnera*, Poaceae) communities related to increased effective precipitation. Analogous communities exist today in the Sub-Antarctic Deciduous Forest in connection with *Sphagnum* bogs and their setting at lower altitudes from southern Isla Grande de Tierra del Fuego where mean annual precipitation totals between 500 and 800 mm and mean annual temperature averages 7 °C (Pisano 1977; Heusser 1998).

Increase in *Empetrum*/Ericaceae type and decrease in *Nothofagus* at ca. 7,500–6,800 cal. yr B.P. suggest a short-term fluctuation to drier conditions. Similar variable moisture conditions were recorded at Puerto Harberton site, in the eastern Beagle Channel (Pendall et al. 2001) and also to the north, at Cerro Frías (SW Argentine Patagonia, 50°S) (Tonello et al. 2009) and Vega Ñandú (southern Chilean Patagonia, 51°S) (Villa-Martínez and Moreno 2007) sites (Fig. 9.1b). These climate conditions were related to a highly variable position and/or intensity of the westerly winds in the southwest of the Andean region between 50–55°S (Villa-Martínez and Moreno 2007; Borromei et al. 2010).

After ca. 6,700 cal. yr BP (subzone IDE-3b), the record of *Drimys winteri*, a typical associate of *Nothofagus betuloides*, along with an increase of *Caltha*, a drop of *Empetrum*/Ericaceae type and record of cushion bog (*Astelia pumila*, *Myrteola nummularia*) and prostate dwarf shrub (*Berberis*, Asteraceae subf. Asteroideae) communities indicate the establishment of Sub-Antarctic Evergreen Forest-Magellanic Moorland transition (Dudley and Crow 1983; Moore 1983). The development of this vegetation unit implies further increase in precipitation that culminated with the establishment and persistence of closed-canopy forest communities of Sub-Antarctic Evergreen Forest between ca. 5,500 and 1,000 cal. yr BP (zone IDE1-4).

The maximum arboreal concentration values recorded at ca. 4,000 cal. yr BP in Isla de los Estados and at ca. 4,500 cal. yr BP in southern Isla Grande de Tierra del Fuego (Beagle Channel area) (Heusser 1989a, 1990, 1998) indicate maximum levels of precipitation and suggest that the core of the westerlies might be focused at 55° S during this time (Ponce et al. 2011; Björck et al. 2012). In Laguna Cascada, the assemblages of diatoms suggest wetter conditions, and a greater development of aquatic vegetation, which would have grown from the edge of the lake from ca. 4,800 cal. years B.P. (Fig. 8.4).

These data correlate well with paleoclimate studies from southern Patagonia at Cerro Frías (Argentina, 50°S; Tonello et al. 2009) and Lago Guanaco (Chile, 51°S; Moreno et al. 2010) pollen records (Fig. 9.1a) and also, within dating uncertainties, with the early onset of Neoglacial (i.e., Holocene) advances in central Patagonia at 46.57°S (Douglas et al. 2005). All this paleoclimate evidence point to northerly latitudinal change and strengthening of the westerlies at these latitudes.

The marine sediment core and Antarctic ice core data showed several sea ice readvances in the course of the Middle to Late Holocene. In particular, the pronounced cooling after ca. 4,000 cal. yr BP was related to the steepening of the summer sea surface temperature (SSST) gradient across the Antarctic Polar Front (APF), strengthening of the westerlies and a northerly shift of the westerly wind belt to its present day latitude (Divine et al. 2010).

The Neoglaciations (=Holocene glaciations) have been recognized in some sectors of the Patagonian Andes (Mercer 1968, 1982; Rabassa et al. 1984) and Fuegian Andes (Coronato 1994; Planas et al. 2001). Although their chronology is still not well defined, the dendrochronological studies (Villalba 1989, 1994) and geomorphological studies (Rabassa et al. 1992) carried out in the Patagonian Andes allow the establishment of the existence of at least five fluctuations during the Holocene related to climatic oscillations.

Mercer (1982) proposed, based on the observed fluctuations in the glaciers of the northeast and east of the South Patagonian Ice (S 48° 20'–S 51° 30'), three glacial advances ("Mercer-type" chronology) at approximately 4,700–4,200 ^{14}C years B.P., 2,700–2,000 ^{14}C years B.P. and the Little Ice Age of the last 300 years (sixteenth To nineteenth Centuries). In the same way, Aniya (1996), based on radiocarbon dates obtained from moraines located in the eastern sector of the South Patagonian Ice, proposes four glacial advances ("Aniya-type" chronology) at approximately 3,600 ^{14}C years B.P., 2,300 ^{14}C years B.P., 1,600–1,400 ^{14}C years B.P and during the Little Ice Age.

The Neoglacial climatic fluctuations have not been clearly identified in the pollen record of the Beagle Channel, although variation in the influx of *Nothofagus* pollen is observed which makes evident the high climatic variability during the last 6,000 cal. years B.P. When the glacial events of the Late Holocene are compared a correspondence can be observed between Neoglacial advances, pulses of greater effective humidity and increase in the influx of *Nothofagus* pollen (Heusser 1998; Moreno et al. 2009).

From ca. 2,700 cal. years B.P. at the west end of Isla de los Estados an important change in the vegetation of the area can be seen. The noticeable reduction in the frequency values and concentration of *Nothofagus* and the absence of *Drimys winteri* (Figs. 7.5 and 7.6) indicate the replacement of the closed rain forest of *Nothofagus* by open forest. The landscape acquires the physiognomy of open ecosystems with herbaceous-bushy vegetation interacting with communities of forest, probably as a consequence of warmer and drier conditions, perhaps by a weakening of the westerlies (Ponce et al. 2011). Similar climate conditions were mirrored in the pollen record from the Cerro Frías site showing a *Nothofagus* forest retraction after 3,000 cal. yr BP and a grass steppe expansion after 800 cal. yr BP related to weakened westerlies (Mancini 2009). According to the pollen data from Lago Guanaco site, the relative opening of the *Nothofagus* woodlands also indicated that precipitation decrease in pulses centered at ca. 2,300–1,300 and 1,000–570 cal. yr BP (Moreno et al. 2009). Toward eastern parts of southern Patagonia such as Laguna Potrok Aike site, located in extra-Andean Patagonia (51°S) (Fig. 9.1a), the pollen record displayed the development of grassy vegetation after ca. 2,300 cal. yr BP suggesting higher moisture availability (Wille et al. 2007) probably due to a stronger easterly moisture influence into the lowlands of southeastern Patagonia by weakening of westerly wind intensities.

In Isla de los Estados, the reduction of the forest reaches the minimum values between ca. 1,000 and 500 cal. years B.P. (Figs. 7.5 and 7.6) in coincidence with the climatic event known as the "Medieval Optimum" or "Medieval Climate Anomaly" (MCA). This event has also been recognized on Isla Grande de Tierra del Fuego, in Lago Fagnano (Fig. 9.1a). The pollen and geochemical information obtained from the study of a sediment core extracted from the lake bed, reveals warm and dry conditions between ca. 1,100 and 574 cal. years B.P. correlated with the climatic anomaly of the "Medieval Optimum" (Borromei et al., in preparation). These conditions were also documented in the Fuegian Andes valleys such as the valley of Andorra (Mauquoy et al. 2004). Further south, the proxy data from Antarctic Peninsula marine cores have registered a period of warmer climate between ca. ~1,200 and 600 cal. yr B.P. equivalent to the "Medieval Warm Period" (MWP) (Bentley et al. 2009).

After 500 cal. years B.P., the recovery of the *Nothofagus* forest indicates colder and more humid conditions (Ponce et al. 2011). This period could be correlated with the event called the "Little Ice Age" (LIA) recorded in the Northern Hemisphere. This event has also been recognized in the Isla Grande de Tierra del Fuego, from pollen records on the peat bogs of some interior Andean valleys, such as the peat bog in Valle de Andorra (180 m a.s.l., Mauquoy et al. 2004) and the peat bog of Las Cotorras, located at 420 m a.s.l. in a high interior valley (Borromei et al. 2010) (Fig. 9.1a).

References

Aniya M (1996) Holocene variations of Ameghino Glacier, southern Patagonia. Holocene 6:247–252

Bentley MJ, Hodgson DA, Smith JA, Cofaigh CO, Domack EW, Larter RD, Roberts SJ, Brachfeld S, Leventer A, Hjort C, Hillenbrand C-D, Evans J (2009) Mechanisms of Holocene palaeoenvironmental change in the Antarctic Peninsula region. Holocene 19(1):51–69

Björck S, Rundgren M, Ljung K, Unkel I, Wallin A (2012) Multy-proxy analysis of a peat bog on Isla de los Estados, easternmost Tierra del Fuego; a unique record of the variable Southern Hemisphere Westerlies since the last deglaciation. Quatern Sci Rev 42:1–14

Borromei AM, Quattrocchio M (2007) Palynology of Holocene marine deposits at Beagle Channel, southern Tierra del Fuego, Argentina. Ameghiniana 41(1):161–171 (Buenos Aires)

Borromei AM, Quattrocchio M (2008) Late and Postglacial Paleoenvironments of Tierra del Fuego: terrestrial and marine palynological evidence. In: Rabassa J (ed) The Late Cenozoic of Patagonia and Tierra del Fuego. Developments in Quaternary Sciences, Elsevier Science Ltd, Vol 11, Chapter 18: 369–381

Borromei AM, Coronato A, Quattrocchio M, Rabassa J, Grill S, Roig C (2007) Late Pleistocene-Holocene environments in Valle Carabajal, Tierra del Fuego, Argentina. J S Am Earth Sci 23:321–355

Borromei AM, Coronato A, Franzén LG, Ponce JF, López Sáez JA, Maidana N, Rabassa J, Candel MS (2010) Multiproxy record of Holocene paleoenvironmental change, Tierra del Fuego, Argentina. Palaeogeogr Palaeoclimatol Palaeoecol 286:1–16

Borromei AM, Waldmann N, Ariztegui D, Olivera D, Martínez MA, Jr. Austin JA, Anselmetti FS (2013) Environmental response to climate oscillations in Tierra del Fuego during the Holocene. (Ms in preparation)

Coronato AMJ (1994) Geomorfología glacial de valles de los Andes Fueguinos y condicionantes físicos para la ocupación humana. Unpublished Doctoral Thesis, Universidad Nacional de Buenos Aires, Facultad de Filosofía y Letras. 318 pp

Coronato AMJ (1995) The last Pleistocene glaciation in tributary valleys of the Beagle Channel. Quaternary of South America & Antarctic Peninsula. Balkema Publishers, Rotterdam, 9:153–172

Coronato AMJ, Salemme M, Rabassa J (1999) Paleoenvironmental conditions during the early peopling of Southernmost South America (Late Glacial-Early Holocene, 14–8 ka BP). Quatern Int 53(54):77–92

Divine DV, Koç N, Isaksson E, Nielsen S, Crosta X, Godtliebsen F (2010) Holocene Antarctic climate variability from ice and marine sediment cores: Insights on ocean–atmosphere interaction. Quatern Sci Rev 29:303–312

Douglass DC, Singer BC, Kaplan MR, Ackert RP, Mickelson DM, Caffee MW (2005) Evidence of Early Holocene glacial advances in southern South America from cosmogenic surface-exposure dating. Geol 33(3):237–240

Dudley TR, Crow GE (1983) A contribution to the Flora and Vegetation of Isla de los Estados (Staaten Island), Tierra del Fuego, Argentina. American Geophysical Union, Antarctic Research Series, Washington, DC 37:1–26

Fernández M (2013) Los paleoambientes de Patagonia Meridional, Tierra del Fuego e Isla de los Estados en los tiempos de las primeras ocupaciones humanas. Estudio basado en el análisis de diatomeas. Unpublished doctoral thesis, Universidad Nacional de La Plata, Facultad de Ciencias Naturales y Museo, La Plata

Gordillo S, Coronato A, Rabassa J (1993) Late Quaternary evolution of a subantarctic paleofjord, Tierra del Fuego. Quatern Sci Rev 12:889–897

Grill S, Borromei AM, Quattrocchio M, Coronato A, Bujalesky G, Rabassa J (2002) Palynological and sedimentological analysis of Recent sediments from Río Varela, Beagle Channel, Tierra del Fuego, Argentina. Revista Española de Micropaleontología 34(2):145–161

Haberzettl T, Corbella H, Fey M, Janssen S, Lücke A, Mayr C, Ohlendorf C, Schäbitz F, Schleser GH, Wille M (2007) Lateglacial and Holocene wet—dry cycles in southern Patagonia: chronology, sedimentology and geochemistry of a lacustrine record from Laguna Potrok Aike, Argentina. Holocene 17(3):297–310

Hajdas I, Bonani G, Moreno PI, Ariztegui D (2003) Precise radiocarbon dating of Late-Glacial cooling in mid-latitude South America. Quatern Res 59:70–78

Hall BL, Porter CT, Denton GH, Lowell TV, Bromley GRM (2013) Extensive recession of Cordillera Darwin glaciers in southernmost South America during Heinrich Stadial 1. Quatern Sci Rev 62:49–55

Heusser CJ (1989) Late Quaternary vegetation and climate of Tierra del Fuego. Quatern Res 31:396–406

Heusser CJ (1990) Late-glacial and Holocene vegetation and climate of subantartic South America. Rev Palaeobot Palynol 65:9–15

Heusser CJ (1998) Deglacial paleoclimate of the American sector of the Southern Ocean: Late Glacial-Holocene records from the latitude of Canal Beagle (55° S). Argentine Tierra del Fuego. Palaeogeogr Palaeoclimatol Palaeoecol 141:277–301

Heusser CJ (2003) Ice Age Southern Andes. A chronicle of paleoecological events. Developments in Quaternary Science 3. In : Rose J (ed) Elsevier, Netherlands 5–10

Heusser CJ, Rabassa J (1987) Cold climatic episode of Younger Dryas Age in Tierra del Fuego. Nat 328(6131):609–611

Isla F, Bujalesky G, Coronato A (1999) Procesos estuáricos en el Canal Beagle, Tierra del Fuego. Revista de la Asociación Geológica Argentina 54(4):307–318

Mancini MV (2009) Holocene vegetation and climate changes from a peat pollen record of the forest-steppe ecotone, Southwest of Patagonia (Argentina). Quatern Sci Rev 28(15–16):1490–1497

Mauquoy D, Blaauw M, van Geel B, Borromei A, Quattrocchio M, Chambers FM, Possnert G (2004) Late Holocene climatic changes in Tierra del Fuego based on multiproxy analyses of peat deposits. Quatern Res 61:148–158

McCulloch RD, Davies SJ (2001) Late glacial and Holocene palaeoenvironmental changes in the central Strait of Magellan, southern Patagonia. Palaeogeogr Palaeoclimatol Palaeoecol 173:143–173

Mercer JH (1968) Variations of some Patagonian glaciers since the Late Glacial: I. Am J Sci 266:91–109

Mercer JH (1982) Holocene glacier variations in southern South America. Striae 18:35–40

Moreno PI, Francois JP, Moy CM, Villa-Martínez R (2010) Covariability of the Southern Westerlies and atmospheric CO_2 during the Holocene. Geol 38(8):727–730

Moreno PI, Francois JP, Villa-Martínez RP, Moy CM (2009) Millennial-scale variability in Southern Hemisphere westerly wind activity over the last 5000 years in SW Patagonia. Quatern Sci Rev 28:25–38

Moore M.D (1983) Flora of Tierra del Fuego. Antony Nelson England, Missouri Botanical Garden, USA, 369 pp

Pendall E, Markgraf V, With JW, Dreier M (2001) Multiproxy record of Late Pleistocene-Holocene climate and vegetation changes from a peat bog in Patagonia. Quatern Res 55:168–178

Pisano E (1977) Fitogeografía de Fuego-Patagonia chilena. Comunidades vegetales entre las latitudes 52° y 56° S. Anales del Instituto de la Patagonia, Punta Arenas 8:121–250

Planas X, Ponsa A, Coronato A, Rabassa J (2001) Geomorphological features of Late Glacial-Holocene Glaciations in Martial Cirque, Fuegian Andes, southernmost South America. Quatern Int 87(1):19–27

Ponce JF, Borromei AM, Rabassa J, Martínez O (2011) Late Quaternary palaeoenvironmental change in western Staaten Island (54.5° S, 64° W), Fuegian Archipelago. Quatern Int 233:89–100

Premoli AC, Mathiasen P, Kitzberger T (2010) Southern-most Nothofagus trees enduring ice ages: Genetic evidence and ecological niche retrodiction reveal high latitude (54° S) glacial refugia. Palaeogeogr Palaeoclimatol Palaeoecol 298(3–4):247–256

Prentice C (1985) Pollen representation, source area and basis size : Toward a unified theory of pollen analysis. Quatern Res 23:76–86

Rabassa J, Brandani A, Boninsegna J, Cobos D (1984) Cronología de la "Pequeña Edad de Hielo" en los glaciares "Río Manso" y "Castaño Overo", Cerro Tronador, Río Negro. Revista de la Asociación Geológica Argentina 41(3–4):405–409

Rabassa J, Heusser C, Rutter N (1990) Late-Glacial and Holocene of Argentine Tierra del Fuego. Quat S Am A penninsula 7:327–351

Rabassa J, Bujalesky G, Meglioli A, Coronato A, Gordillo S, Roig C, Salemme M (1992) The Quaternary of Tierra del Fuego, Argentina: the status of our knowledge. Sveriges Geologiska Undersökning, Ser Ca 81:249–256

Rabassa J, Coronato A, Bujalesky G, Roig C, Salemme M, Meglioli A, Heusser C, Gordillo S, Roig Juñent F, Borromei A, Quattrocchio M (2000) Quaternary of Tierra del Fuego, Southernmost South America: an updated review. Quatern Int 68–71:217–240

Sugden DE, Bentley MJ, Fogwill CJ, Hulton NRJ, McCulloch RD, Purves RS (2005) Late Glacial glacier events in southernmost South America: a blend of "Northern" and "Southern" Hemispheric climatic signals? Geografiska Annaler, (Series A). Phys Geogr 87:273–288

Tonello MS, Mancini MV, Seppä H (2009) Quantitative reconstruction of Holocene precipitation changes in southern Patagonia. Quatern Res 72:410–420

Unkel I, Björck S, Wohlfarth B (2008) Deglacial environmental changes on Isla de los Estados (54.4° S), southeastern Tierra del Fuego. Quatern Sci Rev 27:1541–1554

Unkel I, Fernández M, Björck S, Kjung K, Wolfarth B (2010) Records of environmental changes during the Holocene from Isla de los Estados (54.4° S), southeastern Tierra del Fuego. Global Planet Change 74:99–113

Villa-Martínez R, Moreno PI (2007) Pollen evidence for variations in the southern margin of the westerly winds in SW Patagonia over the last 12,600 years. Quatern Res 68:400–409

Villalba R (1989) Latitude of surface high pressure belt over western South America during the last 500 years inferred from tree-ring analysis. Quat S Am A Penninsula 7:273–303

Villalba R (1994) Tree-ring and glacial evidence for the Medieval Warm Epoch and the Little Ice Age in Southern South America. Clim Chang 26:183–197

Wille M, Maidana N, Schäbitz F, Fey M, Haberzettl T, Janssen S, Lücke A, Mayr C, Ohlendorf C, Schleser G, Zolitschka B (2007) Vegetation and Climate dynamics in Southern South America: the microfossil record of Laguna Potrok Aike, Santa Cruz, Argentina. Rev Paleobot Palynol 146(1–4):234–246

Chapter 10
Archaeology

Abstract The present chapter explores the palaeoenvironmental situation of Isla de los Estados and its relationship with the archaeological record left behind by ancient canoeing people. Was there an environmental barrier to limit the accessibility to the outer edge groups of islands as, for example, Isla de los Estados or Bayly Island (located near Cape Horn)? The palaeoenvironmental information could be used as an explanation for the absence of cultural remains and also for the scarce archaeological sites. The paleoenvironmental conditions inferred from different proxy analysis demonstrate that the Middle Holocene was a windy period, with also higher precipitation rates and variable temperature. The strong wind drifts or the westerlies might have dominated the climatic scenario of the period; hence the navigation conditions would have been almost risky or even impossible. During the late Holocene, milder conditions might have favored seasonal travelling to the outer islands of the Fuegian Archipelago, though no definitive or permanent settlements have been found yet.

Keywords Archaeology · Palaeoenvironment · Fueguian Archipelago · Isla de los Estados · Tierra del Fuego · Human exploration · Late Holocene

10.1 Introduction

One of the aims of this chapter was to analyze why there is no evidence of human occupation at Isla de los Estados during the Middle Holocene, and to discuss if this could be related to certain paleoenvironmental conditions acting in this area and the adjacent seas. This chapter is based on a paper by Fernández et al. (2012). A transcription of the most important aspects of that work is herein presented.

The archaeological sites recorded in Isla de los Estados are BC I and BC II (Chapman 1987), in Crossley Bay, Flinder III (Chapman 1987) and Colnett Beach (Horwitz 1985) (Fig. 10.1b). The first site of this list has the earliest available

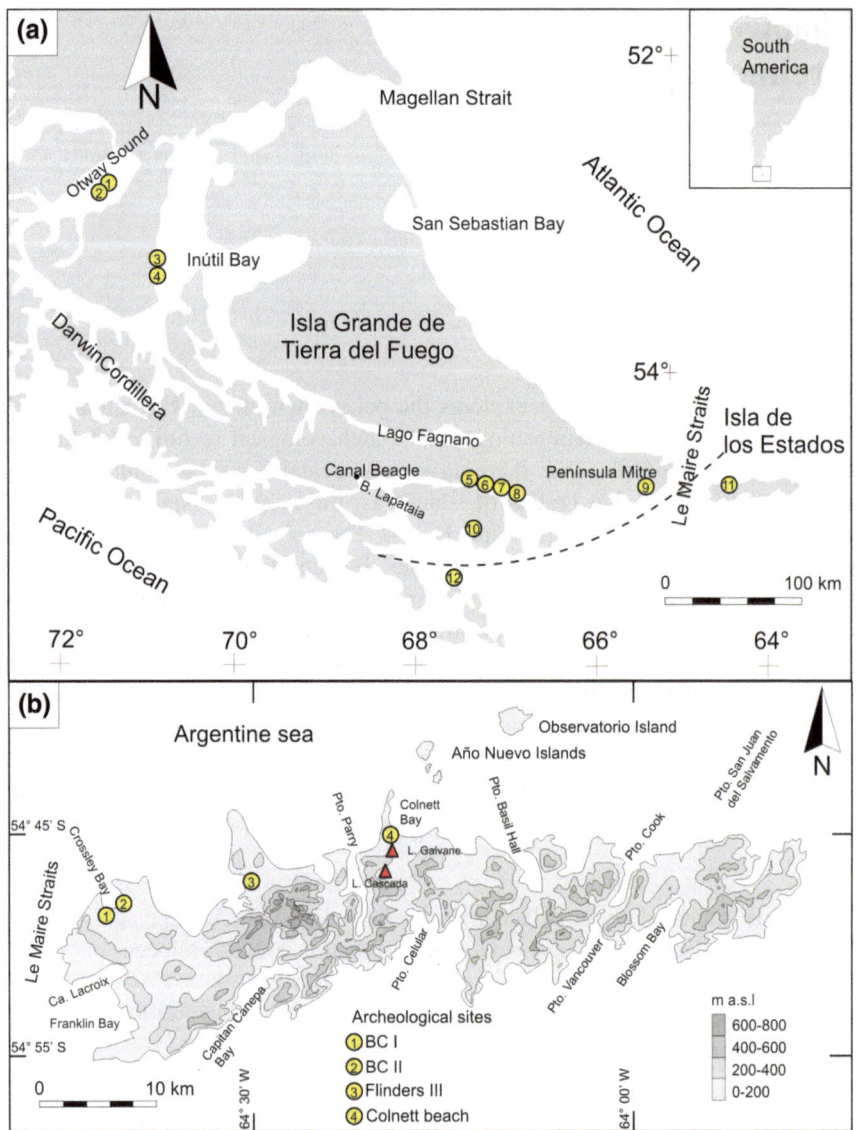

Fig. 10.1 a Fuegian Archipelago. Main archaeological sites during the Middle Holocene are indicated. The *dotted line* points the boundary between some of the most important occupations in the area whose archaeological remains dated from the Late Holocene. *1* Englefield I (6,100 ± 110 ^{14}C years B.P.) (Legoupil 1988); *2* Bahía Colorada (5,900 ± 65 ^{14}C years B.P.) (Legoupil 1997); *3* Bahía Buena (5,895 ± 65 ^{14}C years B.P.) (Ortiz-Troncoso 1975); *4* Punta Santa Ana (6,810 ± 70 ^{14}C years B.P.) (Legoupil and Fotugne 1997); *5* Lancha Packewaia (4,980 ± 70 ^{14}C years B.P.) (Orquera and Piana 1999); *6* Túnel I (6,980 ± 110 ^{14}C years B.P.); *7* Mischiuen I (4,890 ± 210 ^{14}C years B.P.) (Piana et al. 2004); *8* Imiwaia I (7,840 ± 50 ^{14}C years B.P.) (Orquera and Piana 2009); *9* Bahía Valentín (5,900 ± 80 ^{14}C years B.P.) (Vidal 1988); *10* Grandi I (6,120 ± 80 ^{14}C years B.P.) (Legoupil 1993–1994); *11* BC I (2,730 ± 90 ^{14}C years B.P.) (Horwitz 1993); *12* Bayly I (1,410 ± 50 ^{14}C years B.P.) (Legoupil 1993–1994). **b** Isla de los Estados (Staaten Island) with the location of the main archaeological sites and the core location for diatom analysis

radiocarbon dates, comprising a sequence that extends from 2,700 to 1,500 ^{14}C years B.P. (Horwitz 1993). This author interpreted that this site was the result of occasional, seasonal occupation by canoe-travelling people around Península Mitre. But, what happened before this occupation? Between ca. 6,000 and 4,000 cal. years B.P. Península Mitre was already inhabited, according to published dates (Vidal 1988; Zangrando et al. 2009). Moreover, Zangrando et al. (2009) noted that the following period, between ca. 4,000 and 2,000 cal. years B.P., lacks information in Península Mitre, probably due to abandonment processes or population replacement. In addition, the population density around that time could have been lower in the Fuegian Archipelago. Thus, a sampling bias could not be discarded because of a better preservation of earlier archaeological sites (Vázquez et al. 2007).

10.2 Peopling of the Beagle Channel and Neighboring Islands

The early peopling of the southern Fuegian Archipelago has been under research since the beginning of the twentieth century or even before (Legoupil 1993–1994; Legoupil and Fotugne 1997; Ortiz-Troncoso 1972; Ocampo and Rivas 2000; Orquera 2005; Orquera and Piana 1999, 2006). Legoupil and Fotugne (1997) observed that the geographical distribution of the occupation in the area had been discontinuous during the last 6,400 years. They considered two phases or stages in the peopling of this southern region: first, the adaptation to littoral environments in certain spaces called "núcleos de base" (base cores) and possible sites of transition where groups adapted to both terrestrial and marine environments inhabited; second, a later expansion in a progressive and concentric movement toward the outer edge islands or "offshore islands". The aforementioned authors believed that the Magellan Strait-Otway Sound area (Fig. 10.1a) and the Beagle Channel could have been these "núcleos de base" places, mostly because of the earliest sites with adaptation to the littoral way of life located there. Nevertheless, they considered that a process of adaptation like that could have started further north and later expanded southwards by highly mobile small canoeing groups (Legoupil and Fotugne 1997, p. 85).

However, according to Orquera and Piana (1999, 2009), the Beagle Channel would have been a less likely alternative space for the origin of a process of adaptation to littoral environments. Orquera and Piana (2005, 2006) believed that there are other regions which would have been more appropriate for the starting of this specialized littoral life, for example, the island of Chiloé (42° 36′ S; 73° 57′ W) or the Otway Sound-Western portion of the Strait of Magellan, both areas in Chile. They have analyzed the cultural material evidence from these places and those coming from central and southern Patagonia.

Several radiocarbon dates coming from archaeological sites all along the coast of the Chilean fjords and the Beagle Channel set the occupations of that area during the Middle Holocene. In the Beagle Channel, the earliest known radiocarbon date comes from the Imiwaia I site (7,840 ± 50 ^{14}C years B.P.; Orquera and Piana 2009). Together with the earliest assemblage from Túnel I site (Fig. 10.1a, 6,680 ± 210 ^{14}C years B.P.; Orquera and Piana 1999), tool kits remind those of the inland hunters (Orquera and Piana 2009). From there, the adaptive strategy seems to has expanded gradually to the off shore islands looking for good places to settle. But around ca. 6,500 cal. years B.P., the Beagle Channel region was already inhabited by people who were exploiting marine resources such as it is shown in the "Second Component" of Túnel I (TISC) and the "Lower Component" of the Imiwaia I site (LCII) (Orquera and Piana 1999, 2009) (Fig. 10.1a). A characteristic of the local tool kit is the occurrence of two distinct harpoon points; one of them is detachable from the hafts which have a cross-shaped base, whereas the other has a long non detachable base and multiple barbs (Orquera and Piana 1999, 2009). The first inhabitants of the archipelago seems to have developed a special technology (harpoons with detachable points) to be used in these littoral environments. They also could have reached the surrounding islands of the Fuegian seas using canoes. Following this idea, arriving at Isla Navarino (at least at 6,120 ± 80 ^{14}C years B.P.; Legoupil 1993–1994), Isla de los Estados and other islands situated near Cape Horn, would have been possible using canoes and/or rafts. According to the available environmental information for ca. 8,000 ^{14}C years B.P., the Beagle Channel, an ancient glacial valley, was flooded by raising marine waters around this time (Rabassa et al. 1986; Gordillo et al. 2005). Therefore, the possibility for these people to become maritime specialized was already established. The most important favorable factors were abundant marine fauna, sea waters protected from both huge and fast Antarctic waves and furious winds, and availability of forested areas where raw materials to make canoes and harpoons hafts were available (Orquera and Piana 1988). The increase of some significant resources for human subsistence as wood and tree bark might have positively influenced the availability of other resources (Zangrando 2009). Mobility becomes an important strategy to exploit other lesser-used patch than to intensify the exploitation of just only one.

The abundant and homogeneously distributed resources made possible a higher mobility in hunter-gatherer societies, which was precisely what was recorded in this area of the southern cone of the Americas (Orquera and Piana 2006). It might then be a relationship between native people, navigation conditions, coastal landscape, resource availability, and unpredictable weather. Due to these factors, it seems of high importance to explore the prevailing paleoenvironmental conditions during the Middle Holocene.

10.3 Archaeological Sites at Isla de los Estados

The first archaeological survey was conducted by Anne Chapman in 1982. Later on, the studies continued in 1985 by Anne Chapman and Victoria Horwitz (Chapman 1987) and by Horwitz (1986) in the following geographical areas:

Crossley, Colnett, Flinders, San Antonio, Puerto Presidente Roca, Basil Hall, Puerto Parry, Cánepa, and Franklin bays (Horwitz 1993) (Fig. 10.1b).

   After the evaluation of the presence of sites and non-sites in terms of their location, Horwitz (1993) suggested that there are multiple factors that determined the human peopling of Isla de los Estados. For example, food resources availability, habitable space, and distance to the nearest landmass are considered to be equally significant determinant factors.

   The archaeological remains in Isla de los Estados (Fig. 10.1b) include three scatters and one stratified site located at Zaratiegui Bay. The sites defined as scatters are: Palet Beach (Crossley Bay), Flinder III (Flinder Bay), and another site in Colnett Bay (Horwitz 1990).

10.3.1 Flinder III

Anne Chapman collected some lithic artifacts lying on the surface at the base of a sand-dune but no test pits were excavated (Horwitz 1990). The cultural remains include three cores, two flakes, two picks, and two flaked pebbles, all made from pyroclastic devitrified rocks. Also, some birds bones scattered were observed on the surface. No charcoal for dating purposes was found (Horwitz 1990).

10.3.2 Colnett Beach

The cultural remains in this scatter are doubtful and consist of a few cobbles flaked or fractured and three partially weathered long bones (Horwitz 1990). Some test pits were dug but no evidence of archaeological remains was found. The lithic remains are made all in coarse-grained igneous rocks, and include three cobbles, fractured in a vertical plane, one cobble with marginal flaking around the entire perimeter and two large primary flakes. No carbon remains for dating were obtained (Horwitz 1990).

10.3.3 BC I (Crossley Bay)

This site is located on the southwestern section of Playa Zaratiegui, a beach in Crossley Bay. It is a stratified shell-midden which lies on a sandy beach (Horwitz 1990). Anne Chapman located the site in 1982. It appeared as a surface scatter of bones, lithic artifacts, and faunal remains. BC I is one large site which was covered in part by a 4 m high sand-dune. Part of this dune is bounding the site at the south, the northern boundary is closer to the sea, the western, and eastern limits were determined by archaeological criteria. Playa Zaratiegui is protected from cold, southern and northern winds but it is exposed to the western winds (Horwitz 1990, p. 181). In the non-eroded part of the site, there are three different overlapping shell-middens. These shell-midden levels are separated by gray sand levels that also include archaeological remains but in a much lesser quantity (Horwitz and Scheinsohn 1996). These levels were radiocarbon dated by Horwitz (1990), as follows: Level II: 2,180 ± 130 ^{14}C years B.P.; Level IV: 2,480 ± 60 ^{14}C years B.P.; Level VI: 2,730 ± 90 ^{14}C years B.P. However, a younger date for Level II is available: 1,527 ± 58 ^{14}C years B.P. (Chapman 1987, pp. 67).

The portion of BC I on the 2.5 m high terrace is a surface scatter of lithic debris, including two side scrapers, three end scrapers, two knives, one small oval-shaped stone with a groove around half of its circumference, and flakes of pyroclastic devitrified rock. Some seal bones in a very poor conservation condition were also obtained (Horwitz 1990). The eroded portion covered most of the site which was not originally buried under the dune. This is an area with scatter lithic flakes and weathered faunal remains (Horwitz 1990). All lithic artifacts of BC I and BC II were manufactured mainly from marine pebbles found along the coast (Horwitz 1990).

10.3.4 BC II (Crossley Bay)

This site was found in an 8 m high beach terrace. The cultural remains are continually eroded by wind action. The BC II site is located about 30 m above present sea level, on the sand dunes behind Playa Palet (Horwitz 1990). Over the years different materials were found: one flaked pebble and one end scraper (Chapman 1987, p. 64); two end scrapers, six flakes; two end scrapers, one percussion stone, and a few flakes (Horwitz 1990). It is located quite far away from the current cormorants and penguins colonies (Horwitz 1990, p.181). According to Horwitz (1990), the characteristics of such occupation remained unknown. Also, no organic material was found for dating.

10.4 Faunal Assemblages from BC I and BC II Sites

The low quantity of tools or artifacts found in these sites is linked with feeding activities as the catch of fish and terrestrial animals. Mainly, three explanations about the lower found artifact density were given. The assemblage is essentially composed of marine faunal remains. The main taxa represented were penguins, cormorants, pinnipeds, albatrosses, wild geese, cetaceans, and mollusks. This is coincident with the faunal archaeological assemblages found at Beagle Channel area during the Late Holocene (Tívoli and Zangrando 2011). Likely, pinnipeds were the main food resource consumed in BC I. Penguins are represented in all levels of BC I. Many penguin species live and breed even today in Isla de los Estados (Horwitz 1990; Schiavini 2000; Liljesthröm et al. 2008). A small rock-hopper colony is located about one hour and a half walking from the BC I site (on Franklin Bay) (Chapman 1987; Horwitz 1990). Albatross is the largest bird in the area. Currently, this bird lives and breeds in the island. Furthermore, cormorants have a colony in a southeasterly direction from the BC I site. Four species of wild geese are found in the island today. In general, the faunal assemblage was badly preserved due to the action of wind and rain and the aggressive sandy environment (J. L. Lanata, in Horwitz 1990).

The quality of archaeological sites is linked with the dominant geomorphological processes involved in site formation. Wind action is one of the most important factors acting upon the stratified archaeological sites at Isla de los Estados. In the area of Bahía Franklin (southwest of the island), there is a dune field. Between these dunes, peat has developed with a basal ^{14}C date of 10,679 ± 62 years B.P. (AA62509; Ponce et al. 2011). The present wind action not only erodes, but also sand spits are born on the beach, and they advance and bury grasslands and surrounding peatlands (Ponce et al. 2011). It is important to note that the wind direction is coincident with the principal axis of the dunes, which is evidence that the wind direction (coming from the west-southwest) might have remained constant during the entire considered period, from the beginning of the Holocene (Ponce 2009).

At the BC I site, the dune continued growing and then partially covered the southern end of the site (Horwitz 1990, pp. 165). Following this, it is hard to establish the date of the site as a whole. Probably, an earlier level with archaeological remains might be buried under the sand. The mentioned age of 2,700 ^{14}C years B.P. should be considered as a minimum age (Luis Borrero, personal communication). However, the published radiocarbon date is helpful to the purpose of build up a new model.

The faunal assemblage shows that the most important food resource during the earliest occupation was composed of pinnipeds, penguins, cetaceans, cormorants, wild geese, and smaller birds. Along the entire occupation (about 2,000 years) penguins became a much more important food resource relative to the pinnipeds, and the consumption of mollusks also increased (Horwitz 1990).

It is not possible to say that hunter-gatherer lived permanently at Isla de los Estados. This is linked with the hypothesis of Borrero (2001) about the difficulty in affirming the colonization of the island. In fact, it is possible to think this island as just a place where people arrived, stopped, and exploited the natural resources for their subsistence, but which was never colonized definitively.

10.5 Archaeology and Paleoenvironmental Scenarios in the Southern Tip of Tierra del Fuego

The interaction between human culture and environment has been an important topic in anthropological literature concerned with understanding human behavioral patterns and cultural change. For this reason, the comparison of the different available proxies is a good approach in order to reconstruct the paleoenvironmental history of Isla de los Estados.

It is curious and interesting to evaluate why, during the Middle Holocene, a special time period when many environmental and landscape modifications took place around the southern end of South America (Salemme and Miotti 2008), the off shore islands of the Fuegian Archipelago displayed a very restricted or almost absent archaeological evidence for the time considered. Was there an environmental barrier to limit the accessibility to the outer edge groups of islands as, for example, Isla de los Estados or Bayly Island (located near Cape Horn)? Why is there evidence of human activity in southern Navarino Island as early as 6120 ± 80 ^{14}C years B.P. (Grandi I site; Legoupil 1993) but they are absent at Isla de los Estados? Shall we consider cultural decisions? The geomorphology of the island is important to evaluate its occupation. Were there optimal places to be settled? Considering the archaeological data herein presented and taking into consideration the observations made by Chapman (1987) and Horwitz (1986, 1990, 1993), it is interesting to explore the dominant environments that might have affected or limited the settlement of the human groups. Isla de los Estados presents some sheltered places to settle but, considering the archaeological record, they are mainly scattered sites. Sheltered places might have been the first sites to be occupied in stable conditions and they could be presenting a delay in the effective peopling of remote areas (Orquera and Piana 2006). Isla de los Estados may have been used as a place for random or seasonal (Ortiz-Troncoso 1972) visits in which canoeing people might have reached the island just to exploit the resources found in penguin and pinniped colonies.

The paleoenvironmental conditions inferred from the diatom analyses together with the geochemical (Unkel et al. 2010) and pollen studies (Ponce 2009; Ponce et al. 2011) demonstrate that the Middle Holocene was a windy period, with also higher precipitation rates and variable temperature (Fig. 10.2). Within this context, it is interesting to consider the evaluation of Horwitz (1993) about the regional maritime pattern, between isolated islands of the Fuegian Archipelago: "Nomadic maritime hunter-gatherers would likely have exploited any potential resource available while navigating from one island to another" (Horwitz 1993, p. 149). The strong wind drifts or the westerlies might have dominated the climatic scenario of the Middle Holocene; hence the navigation conditions would have been almost impossible. However, it has been said (Horwitz 1990) that weather conditions might have been well-known by the prehistoric groups, and probably they based their "traveling" on a general climatic trend of particular weather situations (moon cycles, windows of good-navigable conditions?). Isla de los Estados was an attractive spot to get to, mainly because of its abundant marine resources. Seasonal expeditions were done to Isla de los Estados and the Cape Horn islands looking for certain resources as penguins and other birds. According to Fitzhugh (1997), the outer places showed a higher instability with a clear period of abandonment and re-occupation in some areas. This is most likely the kind of archaeological evidence found at the island.

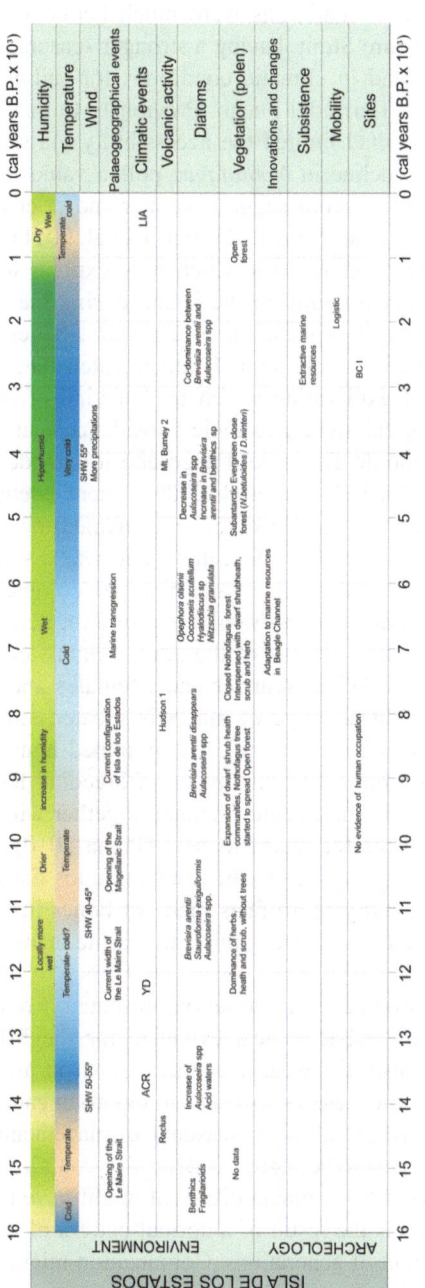

Fig. 10.2 Paleoenvironmental changes in Isla de los Estados based on different proxy data and geoarchaeological information

Following our results, the paleoenvironmental scenarios to reach the off-shore islands was difficult and even risky. Perhaps, after 2,700 ^{14}C yr B.P. (ca. 2,868 ± 87 cal. years B.P, the earliest human occupation found in Isla de los Estados), the environmental conditions were much better to sail and cross the rough waters of the Le Maire Strait, having a stronger seasonality in weather prediction, evaluating the trend in climatic conditions for sailing around the outer margin of Isla Grande de Tierra del Fuego. Previously to that time, the intensity of the southern westerlies had limited the accessibility via sea-navigation mode (Fig. 10.1a). The abrupt decline in *Nothofagus* pollen values after ca. 2,700 cal. years B.P. at Bahía Franklin record suggest warmer and drier conditions perhaps by a weakening of the westerlies (Fig. 10.2) (Ponce et al. 2011).

How we can evaluate other reasons which may explain why the prehistoric groups could not settle occasionally in the island during the Middle Holocene? Could be instead a symbolic explanation or a cultural belief about not exploring the island? Some other factors should not be ruled out, such as: sampling biases, specific events of exploration which in that kind of environment would leave a weak archaeological signal, tectonic subsidence that is recorded in the eastern end of Isla Grande de Tierra del Fuego and it extended up to Isla de los Estados (Rabassa et al. 2003; Ponce et al. 2009). There is ethnographic knowledge that the selk'nam and yamana peoples (who named the Isla de los Estados as "Chuanisin": land of plenty) knew that the island was a special and abundant land with plenty of resources as penguin and pinniped colonies (Chapman 1987, p. 7). The main *Arctocephalus australis* colonies were located on the outer arc of islands and Isla de los Estados. Most of the adult animals concentrate on these islands during the summer. When summer ends, females and offsprings remain close by, but the geographical range of males is greater, since they go farther away, for instance to the Beagle Channel. The pinniped hunting could last with time because humans did not exploit the place of breeding located in the external part of their camp sites. The Beagle Channel, together with Cape Horn, were some of the common places for males to feed (Piana 2010). It is important to mention that the archaeological sites located in Isla de los Estados and Península Mitre shared a higher quantity of artifacts made of bones instead of lithics, and also a more intense exploitation of penguins and marine mammals than of terrestrial fauna (Horwitz 1990).

Overall, the paleoenvironmental information presented here could be an explanation of the absence of archaeological evidence for the Middle Holocene in Isla de los Estados and also, the weaker cultural signal in a closer area such as Península Mitre. During the Late Holocene, the environmental conditions might have been milder (Fig. 10.2), allowing seasonal displacements by canoe to the outer islands of the Archipelago but not their definitive colonization.

According to Borrero (2001), human dispersal might have been slow all along Patagonia. The expansion is the result of a normal pattern of mobility to the long extension of a large amount of generations. Colonization might be found in places where accessibility was simple, with large amount of resources and easy to exploit (considering time and energy consumption) and where settling a camp was easy.

References

Borrero LA (2001) El poblamiento de la Patagonia, Toldos. Milodones y Volcanes. Buenos Aires, Editorial Emecé

Chapman A (1987) La Isla de los Estados en la Prehistoria. Editorial Universitaria de Buenos Aires (Eudeba), Buenos Aires, Primeros datos arqueológicos. Colección Temas

Fernández M, Maidana NI, Rabassa J (2012) Palaeoenvironmental conditions during the Middle Holocene at Isla de los Estados (Tierra del Fuego, 54°, Argentina) and their influence on the possibilities for human exploration. Quatern Int 256:78–87

Fitzhugh W (1997) Biogeographical archaeology in the Eastern North American Arctic. Hum Ecol 25(3):385–418

Gordillo S, Coronato A, Rabassa J (2005) Quaternary molluscan faunas from the island of tierra del fuego after the last glacial maximum. Sci Mar 69:337–348

Horwitz VD (1985) Informe arqueológico de la campaña a las Bahías Colnett, Basill Hall y Puerto Presidente Roca. Unpublished files at the Museo Territorial, Ushuaia, Tierra del Fuego

Horwitz VD (1986) Final report. Isla de los Estados, Bahía Crossley, Argentina. Unpublished files at the Museo Territorial, Ushuaia, Tierra del Fuego

Horwitz VD (1990) Maritime settlements patterns in southeastern Tierra del Fuego (Argentina). Unpublished Ph.D. Dissertation, University of Kentucky, Lexington

Horwitz VD (1993) Maritime settlements patterns: the case from Isla de los Estados (Staaten Island). In: Lanata JL (ed), Explotación de Recursos Faunísticos en sistemas Adaptativos Americanos. Arqueología Contemporánea, Buenos Aires, 4:149–161

Horwitz V, Scheinsohn V (1996) Los instrumentos óseos del sitio Bahía Crossley I (Isla de los Estados). Comparación con otros conjuntos de la Isla Grande de Tierra del Fuego. In: Gómez-Otero J (ed), Arqueología, Sólo Patagonia, CENPAT-CONICET, Puerto Madryn, Argentina, pp 359–368

Legoupil D (1988) Últimas consideraciones sobre las dataciones del sitio de Isla Englefield (seno de Otway). Anales del Instituto de la Patagonia 18, 95–98. Punta Arenas

Legoupil D (1993–1994) El archipiélago del Cabo de Hornos y la costa sur de la Isla Navarino: poblamiento y los modelos económicos. Anales Inst Patagonia 22: 101–121. Punta Arenas

Legoupil D (1997) Bahía Colorada Île Englefield. Recherche sur les Civilisations, Paris

Legoupil D, Fotugne M (1997) El poblamiento marítimo de los archipiélagos de Patagonia: núcleos antiguos y dispersión reciente. Anales Inst Patagonia, Punta Arenas

Liljesthröm M, Emslie SD, Frierson D, Schiavini A (2008) Avian predation at a southern rockhopper penguin colony on Staten Island. Argent Polar Biol 31(4):465–474

Ocampo C, Rivas P (2000) Nuevos fechados ^{14}C de la costa Norte de la Isla Navarino, costa sur del Canal Beagle, Provincia Antártica chilena, Región de Magallanes. Anales Inst Patagonia 28:197–214 Punta Arenas

Orquera LA (2005) Mid-Holocene littoral adaptation at the southern end of South America. Quatern Int 132:107–115

Orquera LA, Piana EL (1988) Human littoral adaptation in the Beagle Channel region: maximum possible age. Quatern S Am Antarct Peninsula 5:133–165

Orquera LA, Piana EL (1999) Arqueología de la región del Canal Beagle (Tierra del Fuego, República Argentina). Soc Argent Antropología, Buenos Aires

Orquera LA, Piana EL (2005) La adaptación al litoral sudamericano sudoccidental: qué es y quiénes, cuándo y dónde se adaptaron. Relaciones Soc Argent Antropología 30:11–32

Orquera LA, Piana EL (2006) El poblamiento inicial del Área Litoral Sudamericana Sudoccidental. Magallania 34(2):21–37

Orquera LA, Piana EL (2009) Sea nomads of the Beagle Channel in southernmost South America: over six thousand years of coastal adaptation and stability. J I Coast Archaeol 4:61–81

Ortiz-Troncoso OR (1972) Nota sobre un yacimiento arqueológico en el archipiélago del Cabo de Hornos. Anales Inst Patagonia 3(1–2):83–85. Punta Arenas

Ortiz-Troncoso OR (1975) Los yacimientos de Punta Santa Ana y Bahía Buena (Patagonia Austral). Excavaciones y fechados radiocarbónicos. Anales del Inst Patagonia 7:93–122. Punta Arenas

Piana EL, Vázquez M, Rua N (2004) Mischiuen I: primeros resultados de una excavación de rescate en la costa norte del canal Beagle. In: Civalero MT, Fernández PM, Guráieb AG (eds) Contra viento y marea. Arqueología de la Patagonia, Instituto Nacional de Antropología y Pensamiento Latinoamericano, Buenos Aires, pp 815–832

Piana EL (2010) Our past and present beliefs on the History of the sea nomads of Tierra del Fuego. Concepts from the 17th to the 21st centuries. In: Del Acebo Ibáñez E, Gunnlaugsson H (eds) La circumpolaridad como fenómeno sociocultural. Pasado, presente y futuro. Facultad de Ciencias Económicas, Universidad de Buenos Aires, pp 217–252

Ponce JF (2009) Palinología y geomorfología del Cenozoico tardío de la Isla de los Estados. Universidad del Sur. Unpublished Doctoral Thesis, Universidad del Sur, Bahía Blanca, Argentina

Ponce JF, Borromei A, Rabassa J, Martínez O (2011) Late Quaternary environmental change in western Isla de los Estados (54°.5 S; 64° W), Fuegian Archipelago. Quatern Int 233:89–100

Ponce JF, Rabassa J, Martínez O (2009) Morfometría y génesis de los fiordos de la Isla de los Estados. Rev Asoc Geol Argentina 65:638–647

Rabassa J, Heusser CJ, Stuckenrath R (1986) New data on Holocene sea transgression in the Beagle Channel (Tierra del Fuego). Quatern S Am Antarct Peninsula 4:291–309

Rabassa J, Coronato A, Roig C, Martínez O, Serrat D (2003) Un bosque sumergido en Bahía Sloggett, Tierra del Fuego, Argentina. Evidencia de actividad neotectónica en el Holoceno Tardío. In: Procesos Geomorfológicos y evolución costera. Actas de la II Reunión de Geomorfología Litoral. Santiago de Compostela, Universidad de Santiago de Compostela, pp 333–345

Salemme M, Miotti L (2008) Archeological Hunter-Gatherer Landscapes Since the Latest Pleistocene in Fuego-Patagonia. In: Rabassa J (eds) Late Cenozoic of Patagonia and Tierra del Fuego. Amsterdam, Elsevier, pp 437–483

Schiavini A (2000) Staaten Island, Tierra del Fuego: the largest breeding ground for Southern Rockhopper Penguins? Waterbirds 23(2):286–291

Tívoli MA, Zangrando FA (2011) Subsistence variations and landscape use among maritime hunter-gatherers. A zooarchaeological analysis from the Beagle Channel (Tierra del Fuego, Argentina). J Archaeol Sci 38:1148–1156

Unkel I, Fernández M, Björck S, Ljung K, Wohlfarth B (2010) Records of environmental changes during the Holocene from Isla de los Estados (54.4° S), southeastern Tierra del Fuego. Glob Planet Change 74:99–113

Vázquez M, Zangrando F, Tessone A, Ceraso A, Sosa L (2007) Arqueología de Bahía Valentín (Península Mitre, Tierra del Fuego): nuevos resultados y perspectivas. In: Morello F et al. (eds) Arqueología de Fuego-Patagonia. Levantando piedras, desenterrando huesos…y develando arcanos. CESQUA, Punta Arenas, pp 755–766

Vidal H (1988) Bahía Valentín: 6000 años de ocupación humanas en el oriente fueguino.In: IX National Congress of Argentine Archaeology, Buenos Aires, p 77

Zangrando FA (2009) Historia evolutiva y subsistencia de cazadores-recolectores marítimos de Tierra del Fuego. Sociedad Argentina de Antropología, Buenos Aires

Zangrando F, Tessone A, Vázquez M (2009) El uso de espacios marginales en el archipiélago fueguino: implicancias de la evidencia arqueológica de Bahía Valentín. In: Salemme M, Santiago F, Álvarez M, Piana E, Vázquez M, Mansur ME (eds) Arqueología de la Patagonia, una mirada desde el último confín. Editorial Utopías, Ushuaia, pp 47–62